"中国森林生物多样性监测网络"丛书　马克平　主编

天童亚热带森林动态样地
——树种及其分布格局

Tiantong Subtropical Forest Dynamics Plot:
Tree Species and Their Distribution Patterns

杨庆松　刘何铭　杨海波　王樟华　方晓峰　马遵平　王希华　著

中国林业出版社
China Forestry Publishing House

图书在版编目 (CIP) 数据

天童亚热带森林动态样地：树种及其分布格局 / 杨庆松等著. —北京：中国林业出版社，2019.10
ISBN 978-7-5219-0350-8

I. ①浙⋯　II. ①杨⋯　III. ①自然保护区－森林植物－介绍－浙江　IV. ①S759.992.55

中国版本图书馆CIP数据核字(2019)第258720号

内容简介

本书介绍了浙江天童常绿阔叶林胸径大于1cm的常见木本植物154种，每个物种以文字简要描述主要分类特征，以彩色照片展示植物的树干、小枝、花、果或幼苗等，并附每个物种在20 hm^2 森林动态样地内的种群空间分布、个体数量和径级结构。同时对该样地的地形、土壤、植被等皆有介绍。本书以资料翔实、图片精美为特色，是亚热带常绿阔叶林不可多得的参考书，也可以作为植物爱好者了解亚热带森林、认识森林植物的野外手册。

中国林业出版社·林业出版分社

策划、责任编辑：于界芬
电　话：010-83143542

出　版	中国林业出版社 (100009　北京西城区德内大街刘海胡同7号)
网　址	http://lycb.forestry.gov.cn/
发　行	中国林业出版社
印　刷	固安县京平诚乾印刷有限公司
版　次	2019年12月第1版
印　次	2019年12月第1次
开　本	889mm×1194mm　1/16
印　张	11.5
字　数	387千字
定　价	148.00元

凡本书出现缺页、倒页、脱页等质量问题，请向出版社图书营销中心调换。
版权所有　侵权必究

序 言 1

在过去的几十年时间里，中国科学院和林业、农业等相关部门陆续建立了数百个生态系统定位研究站。其中，中国科学院组建的中国生态系统研究网络 (CERN) 拥有分布于全国包括农田、森林、草地、湿地、荒漠等生态系统类型的36个生态站。国家林业局建立的中国森林生态系统研究网络 (CFERN) 由29个生态站组成，基本覆盖了我国典型的地带性森林生态系统类型和最主要的次生林、人工林类型。

随着研究的发展，特别是近年来人们对生物多样性和全球变化研究的关注，国际上正在推动生态系统综合研究网络平台的建立。在全球水平上，全球生物多样性综合观测网络 (GEO BON) 是一个有代表性的研究网络。它试图把全球与生态系统和生物多样性长期定位研究相关的网络整合起来，通过综合研究，探讨生态系统与生物多样性维持与变化机制以及系统之间的相互作用机理，为生态系统可持续管理与生物多样性的保护提供科学依据和管理模式。

近年来，中国科学院生物多样性委员会组织建立了中国森林生物多样性监测网络 (Chinese Forest Biodiversity Monitoring Network，以下简称CForBio)。中国是生物多样性特别丰富的少数国家之一，也是唯一一个具有从北部寒温带到南部热带完整气候带谱的国家。截止到2018年，中国森林生物多样性监测网络包括大型监测样地17个，成为继美国史密森研究院热带研究所建立的全球森林生物多样性监测网络 (CTFS-Forest GEO) 之后又一大型区域监测网络。由于CForBio横跨多个纬度梯度，对于揭示中国森林生物多样性形成和维持机制，以及森林生物多样性对全球变化的响应，科学利用和有效保护中国森林生物多样性资源具有重要意义。

目前，CForBio已经有很好的研究进展，各样地研究成果陆续在国际著名生态学刊物如*Ecology Letters*, *Journal of Ecology*等上发表，受到国内外同行的高度评价。但这些文章都是关于某一具体问题的研究总结，还无法让国内外同行全面了解CForBio各个样地整体情况。因此，出版这套以中英文形式介绍各大样地基本情况的"中国森林生物多样性监测网络"丛书是非常必要的。感谢马克平研究员组织相关专家编写这套丛书。我相信该丛书不仅是国内外同行深入了解CForBio各样地的参考书，同时也将为我国森林生物多样性监测和森林生态系统联网研究奠定重要的基础。

（孙鸿烈）
中国科学院前副院长

Foreword 1

In the past few decades, hundreds of Ecosystem Research Stations have been set up by the Chinese Academy of Sciences, State Forestry Administration, Ministry of Agriculture and other relative departments. Among them, 36 ecological research stations were established by Chinese Ecosystem Research Network (CERN), supported by the Chinese Academy of Sciences. The 36 research stations are scattered over the country representing diverse ecosystems, including farmland, forest, grassland, wetland, desert and others. Moreover, the Chinese National Ecological Research Network (CFERN), supported by the State Forestry Administration, consists of 29 research stations, covering typical zonal forest ecosystems and main secondary forests and plantations in China.

With the development of research, especially the growing concern over researches on biodiversity and global change in recent years, the establishment of ecosystem research network have been promoted under international supports. So the Group on Earth Observations Biodiversity Observation Network (GEO-BON) is representative across the world, and it attempts to integrate worldwide networks relating to long-term research on ecosystem and biodiversity. Based on the comprehensive studies, the maintenance and change mechanism of ecosystem and biodiversity and their interactions have been explored, which provide scientific basis and management mode for sustainable development of ecosystem and protection of biodiversity.

In recent years, Biodiversity Committee of the Chinese Academy of Sciences has organized and established Chinese Forest Biodiversity Monitoring Network (CForBio). Chinese is among the few countries with particularly rich biodiversity, and the only country with the whole climate spectrum from north cold temperate zone to south tropical. Till 2018, CForBio consists of 17 large scale monitoring plots, and it is the second large scale regional monitoring network after ForestGEO organized by the Smithsonian Tropical Research Institute. Due to its covering several latitude gradient, CForBio is significant for revealing formation and maintenance mechanisms of Chinese forest biodiversity, responses of biodiversity to global change, scientific utilization and effective protection of Chinese forest biodiversity resources.

Encouraging progress has been made in this area since the network built, for lots of research findings have been published in the international peer reviewed ecological journals, such as *Ecology Letters, Journal of Ecology* and *Oikos,* etc., which brought about positive response from colleagues in the field of plant ecology. However, the published papers mostly focus on research of specific problems; scientists and public still can't understand the whole situation of each plot in details. So it is really necessary to publish this series, which introduce basic information of permanent forest plots in both Chinese and English. I am grateful to Professor Keping Ma for organizing related specialists to prepare the series. And I believe that this series would be a valuable reference book for scientists and public to further understand CForBio, and it will also lay a foundation for the forest biodiversity monitoring and forest ecosystem research in China.

Honglie Sun
The former Vice-President for the Chinese Academy of Sciences

序 言 2

森林在维持世界气候与水文循环中起着根本性的作用。森林是极为丰富多样的动物、植物与微生物的家园，而人类正是依靠这些生物获取各种产品，包括食品与药物。尽管对人类福祉如此重要，森林仍然遭受着来自土地利用与全球气候环境变化的巨大威胁。在这种不断变化的情况下，为了更好地管理全球剩余的森林，迫切需要树种在生长、死亡与更新方面的详细信息。

中国森林生物多样性监测网络 (CForBio) 正在中国沿着纬度与环境梯度建立大尺度森林监测样地。通过这个重要的全国行动倡议与来自中国科学院及若干其他单位的研究者的努力，CForBio开始搜集关于中国森林的结构与动态的关键信息。现在CForBio与史密森研究院及哈佛大学阿诺德树木园的全球森林监测网络 (ForestGEO) 形成了合作伙伴。ForestGEO是个在27个热带或温带国家拥有长期大尺度森林动态研究样地的全球性网络。CForBio与ForestGEO合作的目标是通过合作研究，了解森林是如何运作的，它们是如何随着时间而改变的，以及如何重建或者恢复，以确保森林提供的环境服务能可持续或者增长。森林及其提供的服务的长期可持续性有赖于我们预测森林对全球变化，包括气候与土地利用变化的响应的能力，以及我们去理解与创建适当的森林服务市场的能力。通过拥有67个森林大样地的全球网络及大量项目的训练与能力建设，CForBio与ForestGEO的伙伴关系是发展这些预测工具的重要基础。这种伙伴关系也将促进为全球各地的当地社区、林业管理者与政策制定者在森林的保育与管理方面发展应用性的林业项目建议，发展与示范利用乡土物种进行森林重建的方法，以及从经济学角度评估森林在减缓气候变化、生物多样性保护和流域保护上的价值的方法。

我祝贺作者们创作了这部关于样地植物的优秀丛书。本丛书为将来的森林监测提供了基准信息，是涉及森林恢复、碳存储、动植物关系、遗传多样性、气候变化、局地与区域保育等研究内容的研究者、学生与森林管理者们有价值的参考资料。

S. J. 戴维斯
主任
史密森热带研究所全球森林监测网络

Foreword 2

Forests play an essential role in regulating of world's climatic and hydrological cycles. They are home to a vast array of animal, plant and microorganism species on which humans depend for many products, including food and medicines. Despite the importance of forests to human welfare they are under enormous threat from changes in land-use and global climatic conditions. In order to better manage the world's remaining forests under these changing conditions detailed information on the dynamics of growth, mortality and recruitment of tree species is urgently needed.

The Chinese Forest Biodiversity Monitoring Network (CForBio) that aims to establish large-scale forest monitoring plots across latitudinal and environmental gradients in China. Through this important national initiative, researchers from the Chinese Academy of Sciences and several other research institutions in China, CForBio has begun to gather key information on the structure and dynamics of China's forests. The CForBio initiative is now partnering with the Forest Global Earth Observatory (ForestGEO) of the Smithsonian Research Institute and the Arnold Arboretum of Harvard University. ForestGEO is a global program of long-term large-scale forest dynamics plots in 27 tropical and temperate countries/areas. The goal of the partnership between CForBio and ForestGEO is to work together to understand how forests work, how they are changing over time, and how they can be re-created or restored to ensure that the environmental services provided by forests are sustained or increased. The long-term sustainability of forests and the services they provide depend on our ability to predict forest responses to global changes, including changes in climate and land-use, and our ability to understand and create appropriate markets for forest services. The CForBio-ForestGEO partnership is ideally poised to develop these predictive tools through a global network of 67 large forest plots and an extensive program of training and capacity building. The partnership will also lead to the development of applied forestry programs that advise local communities, forest managers and policy makers around the world on conservation and management of forests, to develop and demonstrate methods of native species reforestation, and to economically value the roles that forests play in climate mitigation, biodiversity conservation, and watershed protection.

I congratulate the authors on the production of this excellent new series of stand books. In addition to providing a baseline for future forest monitoring, these books provide a valuable resource for researchers, students, and forest managers dealing with issues of forest restoration, carbon storage, plant-animal interactions, genetic diversity, climate change, and local and regional conservation issues.

Stuart Davies

Director
Forest Global Earth Observatory / ForestGEO
The Smithsonian Tropical Research Insitute

前 言

常绿阔叶林分布在我国湿润亚热带地区，是中国植被中的一个独特类型。相比于北半球中纬度的其他地区，由于副热带高压和热带气旋的影响，多分布的是荒漠和半荒漠，而我国这一纬度，由于季风气候的影响，加上三级台地独特的地形条件，大面积分布的是常绿阔叶林。常绿阔叶林被科学界认为是重要的物种起源中心，生物多样性极为丰富，是人类获取生物资源的重要基因库。

天童国家森林公园位于浙江省宁波市鄞州区东部的太白山东麓，我国著名五大丛林寺庙之一的天童寺坐落其中。该寺开创于西晋永康元年(公元300年)，迄今已有1700多年的历史，素有"东南佛国"的称号，也是日本佛教曹洞宗的发源地。千百年来，寺庙得森林荫庇香火旺盛，森林也因寺庙得以保护，该地保存有完好的地带性常绿阔叶林，这在我国东部沿海交通便利、人口密集的地区十分难得。1983年华东师范大学宋永昌先生带领团队开始在天童国家森林公园内开展常绿阔叶林研究，并在1986年将此地作为华东师范大学环境科学系生态学野外实习基地，1992年，在天童国家森林公园和天童林场的支持下正式建立了华东师范大学天童生态实验站，经过多年的持续建设，2005年，天童站加入国家生态系统观测研究网络(CNERN)，成为首批国家野外台站成员，是隶属教育部的四个国家野外站之一，也是当时华东地区唯一的森林生态系统国家野外科学观测研究站。

初次知晓大型森林动态监测样地的概念是2001年在加拿大维多利亚大学进修期间。当时，与在维多利亚加拿大国家太平洋林业中心工作的何芳良教授有多次深入交流，也从他那里了解到美国Smithsonian研究院热带森林科学研究中心(Center for Tropical Forest Science，CTFS；现为全球森林监测网络，Forest Global Earth Observatory，ForestGEO)在全球开展的大型森林动态监测样地建设。CTFS建立了标准化、长周期的植物及环境监测体系，通过在大型森林样地中监测植物的种子散布、更新、生长、死亡过程以及树木空间信息和环境因子信息，为探讨群落水平上生物多样性的维持机制提供了重要的基础研究平台。有所了解后，彼时已萌生在天童按照CTFS标准建立大型监测样地的想法。

2003年，何芳良教授就在国内倡导森林动态监测样地的建设，在马克平研究员和何芳良教授的共同推动下，2004年2月，在北京召开的CTFS Mapping Plots-Beijing Workshop中，决定首批建设5个样地，天童是其中之一，也是当时唯一隶属于高校的样地。在中国科学院的支持下，2004年隶属于中国科学院的4个样地相继开始建设。相比于科学院系统，高校建设如此大型的森林动态监测样地，在人力和财力上都有很多困难，筹建过程也历时较久。通过多年努力，积攒资金，天童样地终于在2008年4月开始建设。

天童样地海拔落差近300 m，地形极为复杂，完成如此大型的样地建设工作凝结了很多人的智慧与汗水。样地建设前，何芳良教授多次到天童考察，在天童国家森林公园的核心区域初步确定了样地的位置。中国科学院植物研究所米湘成博士和任海保博士协助完成了样地基线标定，正式确定了样地位置。其后，样地建设先后进行了样地标定及地形测绘 (2008年4月至2009年5月)、每木调查 (2009年7月至2010年10月) 两个阶段。除本书的作者外，参与样地建设工作的还有课题组郑泽梅老师，研究生王达力、张志国、李萍、谢玉彬、张娜、丁慧明、陈静静、吕妍、孟祥娇；本校研究生和本科生戚裕锋、王嵘、熊申展、陈超、张伟军、冯圣铠、牛立峰、官布策仁那木德格、于超、郑蓉、周陶冶、许靖梓、王依云、卫以琦、吴姮、潘亚卉、白颖、韦海、陆皆宇、程亮、王一舟、岳春光、邱磊、朱文杰、王文国、胡海群、朱薛伟、王仡桐、周维、孙向朋；加拿大阿尔伯塔大学阮晓峰，天目山保护区管理局牛晓玲；合肥市撮镇中学吴俊伟。样地建设历时2年半，共计测量树木10万余株。样地调查完成后，部分样地物种的鉴定得到了浙江师范大学陈建华老师帮助。2014年10月，样地开始首次复查并校对两次调查数据，参与复查工作的是课题组研究生费希旸、董舒、袁铭皎、孙小颖、曹烨、苏渝钦、季德成、邢九州、林庆凯、宗意、林莉、李彬彬、阿尔达克·阿庆、任思远、孙小伟、薛生桂。整个建设过程中，宁波市天童林场王良衍场长、王阿昌场长、王波技术员为野外调查提供了工作和生活便利。

　　本书是对天童样地建设成果的初步总结，在介绍区域植被概况、样地群落结构的同时，详尽描述了20 hm^{2}样地中154个树种的物种特征、分布格局和径级结构，并配以彩色照片呈现，为今后基于天童样地平台开展各类研究工作提供了必要的信息。

　　在本书编纂过程中，上海市辰山植物园汪远、严靖、葛斌杰、杜诚、徐晔春提供了部分精美照片；书稿完成后承蒙张健教授和沈国春副教授协助审校，中国林业出版社于界芬女士精心编辑，在此表示诚挚的谢意。

王希华
2019年9月

Preface

Evergreen broad-leaved forest (EBLF) is a unique Chinese vegetation type, distributing in humid subtropical region of China. Due to the impact of subtropical high and tropical cyclone, desert and semi-desert vegetations are prevalent within areas in mid-latitudes of the Northern Hemisphere. Under the combined effects of monsoon climate and the special topographic conditions, areas at the same latitudes of China are largely covered by EBLFs. EBLFs are rich in biodiversity and deemed as a crucial center for species origin. They are the important genetic pools of biotic resources as well.

Tiantong National Forest Park (TNFP) lies on the east side of Taibai Mountain, located in the Eastern part of Yingzhou District, Ningbo City. Tiantong Temple, one of the five famous Jungle Buddhist temples in China, situated within the park. It was first established 1700 years ago, i.e., in the first year of the age of Yongkang of Western Jin Dynasty (300 AD). The temple, known as the title of "Southeastern Buddhism", is the birthplace of Soto sect of Japanese Buddhist. Innumerable pilgrims have come to worship Buddha in Tiantong temple surrounded by the forest for millennia, and therefore the forest was protected by the temple. The well preserved zonal EBLF is rather rare in the eastern coastal areas with dense population and convenient transportation. The team led by Prof. Yongchang Song from East China Normal University (ECNU) began to conduct their research on the EBLF within TNFP in 1983. They subsequently established an ecological field practice base for department of environmental science of ECNU in 1986. The ecological experiment station of ECNU, namely Tiantong experiment station, was officially established with the support of TNFP and Tiantong Forest Farm in1992. After several years of development, the Tiantong experiment station was chosen as the first members of the Chinese National Ecosystem Research Network (CNERN) in 2005. It was one of the four national filed stations belonging to the Ministry of Education, and the only national forest ecosystem observation and research station in East China.

The first time I heard about the concept of Forest Dynamic Plot (FDP) was when I attended advanced studies at University of Victoria. Through the extensive contacts with Prof. Fangliang He who worked in Pacific Forestry Centre of Canada, then I knew the Center for Tropical Forest Science (CTFS), nowadays known as the Forest Global Earth Observatory (ForestGEO), had carried out the construction of FDPs around the world. ForestGEO created standardized and long-term monitoring protocols on plant and environment. Researchers are able to monitor the dispersal-regeneration-growth-death process of plant individuals, spatial information of trees and various environmental elements within an FDP, which enable them to explore the mechanisms of biodiversity maintenance at community-scale. I, to be exact, came up with the idea of establishing an FDP in Tiantong following the CTFS protocols at that time.

Prof. Fangliang He started to promote the FDP construction in China from 2003. With the joint promotion by him and Prof. Keping Ma, it was determined to establish five FDPs first during the CTFS Mapping plots-Beijing workshop in Feb. 2004, including Tiantong FDP - the only FDP belonging to a university at that time. However, it was difficult in both human and financial resources for a university to establish a 20-hm^2 FDP, and the preparation process would last for a long time. The construction of Tiantong FDP was started in Apr. 2008 after years of hard work and adequate fund accumulation.

The altitude difference of Tiantong FDP is approximately 300 m, and its terrain is extremely complex. Therefore, it was a hard work to complete the large construction project. Prof. Fangliang He has visited Tiantong several times before the construction, and preliminarily determined the location of FDP within the core area of TNFP. Dr. Xiangcheng Mi and Dr. Haibao Ren, from institute of botany, the Chinese Academy of Sciences, help us setting the baseline of Tiantong FDP and determining its location formally. The establishment process could be divided into two stages: grid locating and topography mapping (Apr. 2008~May. 2009), and tree census (Jul. 2009~Oct. 2010). Besides the authors of this book, Prof. Zemei Zheng and a large number of students and teachers contributed to the construction of Tiantong FDP, including students Dali Wang, Zhiguo Zhang, Ping Li, Yubin Xie, Na Zhang, Huiming Ding, Jingjing Chen, Yan Lv, Xiangjiao Meng, Yufeng Qi, Rong Wang, Shenzhan Xiong, Chao Chen, Weijun Zhang, Shengkai Feng, Lifeng Niu, GonpoTsering Lhamo Dugkar, Chao Yu, Rong Zheng, Taoye Zhou, Jingzi Xu, Yiyu Wang, Yiqi Wei, Heng Wu, Yhui Pan, Ying Bai, Haijun Wei, Jieyu Lu, Liang Cheng, Yizhou Wang, Chunguang Yue, Lei Qiu, Wenjie Zhu, Wenguo Wang, Haiqun Hu, Xuewei Zhu, Yitong Wang, Wei Zhou, Xiangpeng Sun, Xiaofeng Ruan (University of Alberta), and Ms. Xiaoling Niu (Tianmu Mountain Nature Reserve Administration), Junwei Wu (Cuozhen Middle School). The establishment of Tiantong FDP lasted for two and a half years. More than 100,000 woody plant individuals were measured. Prof. Jianhua Chen of Zhejiang Normal University gave aid to species identification after the vegetation survey. In Oct. 2014, we recensused Tiantong FDP and checked the census data in these two rounds. Master students in our research group participated in the second-round census, including Xiyang Fei, Shu Dong, Mingjiao Yuan, Xiaoying Sun, Ye Cao, Yuqin Su, Decheng Ji, Jiuzhou Xing, Qingkai Lin, Yi Zong, Li Lin, Binbin Li, Ardak·Aqing, Siyuan Ren, Xiaowei Sun, Shenggui Xue. Through the construction of Tiantong FDP, Liangyan Wang, Achang Wang and Bo Wang gave great support to our work and life.

This book preliminarily summarized the construction achievement of Tiantong FDP. After overviewing the zonal vegetation and community structure, we described the characteristics, distribution patterns and DBH structures of the 154 woody plants within the 20-hm^2 plot in detail. Numerous species pictures were attached to the species introduction section. We hope the book can provide some necessary information for the future research works in Tiantong FDP.

While preparing the book, Yuan Wang, Jing Yan, Binjie Ge, Cheng Du and Yechun Xu, from Shanghai Chenshan Plant Science Research Center, provided abundant beautiful pictures of plants. We are delighted to acknowledge the assistance of revision by Prof. Jian Zhang and Associate Profesor Guochun Shen. In the production stage, we are deeply grateful for the careful editing by Ms. Jiefen Yu of China Forestry Publishing House Press.

Xihua Wang
2019.9

目 录

1 天童国家森林公园 ·· 1
   1.1 地理位置和自然环境 ··· 2
   1.2 主要植被类型 ··· 3

2 天童亚热带常绿阔叶林20 hm² 森林样地 ·· 7
   2.1 样地概况 ·· 8
   2.2 样地建设和植被调查 ··· 8
   2.3 地形与土壤 ··· 8
   2.4 物种组成与群落结构 ··· 8

3 树种及其空间分布 ·· 11

附录I 植物中文名索引 ·· 166
附录II 植物拉丁名索引 ··· 168

Contents

1 Introduction to Tiantong National Forest Park ·· 1
   1.1 Location and Natural Environment ·· 2
   1.2 Main Vegetation Types ·· 4

2 The 20 hm² Subtropical Broad-leaved Evergreen Forest Plot in Tiantong ············ 7
   2.1 Overview of the Plot ··· 8
   2.2 Plot Establishment and Vegetation Sampling ····································· 8
   2.3 Topography and Soil ·· 8
   2.4 Species composition and Community structure ··································· 9

3 Tree Species and their Spatial Distribution ·· 11

Appendix I Chinese Species Name Index ·· 166
Appendix II Scientific Species Name Index ··· 168

天童国家森林公园
Introduction to Tiantong National Forest Park

I

1.1 地理位置和自然环境

天童国家森林公园位于浙江省宁波市鄞州区东南部 (图1-1),距宁波市区28 km,地处北纬29°48′,东经121°47′,面积349 hm²。东南名刹天童寺坐落其中,该寺开创于西晋永康元年,迄今已有1700多年的历史。

该地区在地质构造上属于华夏陆台范围,位于闽浙地盾北部,为四明山、天台山的余脉,处于浙东丘陵与滨海平原的交错地带。第四纪以来,本区一直表现为下降运动。公园所在地形形似座椅,三面环山,南向一面宽谷,主峰太白山海拔653.3 m,一般山峰海拔300 m左右。坡度多在10°~30°之间,很少有45°以上山坡[1]。

该区域气候为温暖湿润的亚热带季风气候,全年温和多雨,四季分明。年平均温度为16.2℃,最热月为7月,平均温度为28.1℃,最冷月为1月,平均温度为4.2℃,无霜期237.8天。年平均降水量为1374.7 mm,多集中在夏季 (6~8月),占全年降水量的35%~40%,冬季 (12月至翌年2月) 冷而干燥,降水量仅占全年的10%~15%,春季降水量一般大于秋季 (9~11月) (表1-1)。因受梅雨锋系和台风影响,年内降水主要有两个高峰,分别出现在5~6月和7~8月。年平均相对湿度82%,且变率不大,各季之间最大变率在5%以下。年蒸发量为1320.1 mm,小于降水量,只有7~10月蒸发量稍大于降水量。总体来看该区域雨水充沛、热量充足、水热同季,有利于植物生长和植被发育[1]。

公园内的土壤主要为山地黄红壤,成土母质主要是中生代的沉积岩和部分酸性火成岩以及石英砂岩和花岗岩残积风化物。土层厚薄不一,一般在1 m左右,质地以中壤至重壤为主,全氮和有机质含量较高,一般在0.2%~0.4%和3%~5%之间,土壤pH值多为4.5~5.0[1]。

1.1 Location and Natural Environment

Tiantong National Forest Park (29°48′N and 121°47′E) is located in the Yinzhou District (Fig.1-1), 28 km south of Ningbo City, Zhejiang Province, covering a total area of 349 hectares. There is a Buddhist temple called "Tiantong Temple" situated within the park. It was first established 1600 years ago, i.e., the first year of the age of Yongkang of Western Jin Dynasty.

This area is the extension of Siming Mountain and Tiantai Mountain. It lies in the east hills and coastal plains interlaced zone, at the north part of the Zhejiang and Fujian shield, geologically belonging to the Cathaysia Block. The regional crust is in subsidence since the Quaternary. The park is surrounded by mountains on three sides, and its south side is a valley, topographically characterized as the seat-shape. Taibai Mountain, the main peak of Tiantong, is about 653.3 m asl, and the average elevation of the remaining peaks is about 300 m. The slope of the region ranges from 10° to 30°, rarely exceeding 45°[1].

The region has a warm and wet subtropical monsoon climate with a hot, humid summer and a drier, cold winter. The annual mean air temperature is 16.2 ℃. The warmest and coldest months are July and January, with 28.1 and 4.2 ℃ monthly mean temperatures, respectively. The frost-free period is about 237 days. Mean annual precipitation is 1374.7 mm, concentrating from June to August, which is responsible for 35% to 40% of the annual precipitation. On the contrary, the proportion is merely 10% to 15 % in winter (December to February). The precipitation is generally greater in spring than in autumn (Table 1-1). Due to the influence of the Meiyu front system and the typhoon, there are two precipitation peaks within a year, respectively from May to June and from July to August. Average annual relative humidity is 82%. Its variability is low, e.g., the maximum variability among seasons is less than 5%. Annual evaporation (1320.1 mm) is less than the annual precipitation, but the evaporation from July to October is slightly higher than the precipitation during that period. Overall, abundant rainfall, adequate heat, and the synchronization of rainfall and heat are beneficial to plant growth and vegetation development in this region[1].

The soils of this area are mainly red and yellow ultisols. The substrate of parental material is composed of Mesozoic sediments and acidic intrusive rocks, including quartzite and granite. Soils are of variable depth and mostly around 1 m in average. Soil texture is mainly medium to heavey load. Concentrations of total nitrogen and organic matter are high, respectively ranging from 0.2% to 0.4% and 3% to 5%. Soil pH values from 4.5 to 5.0[1].

表1-1 天童国家森林公园附近气象站的气象记录（1953~1980，鄞州区气象站）
Table 1-1 Records of the meterological station at Yinzhou District in Ningbo nearby Tiantong National Forest Park (1953~1980, Yinzhou District Meterological Station)

月份 Month	气温 Temperature (℃)	降水量 Precipitation (mm)	相对湿度 Relative humidity (%)	蒸发量 Evaporation (mm)	日照时数 Sunshine hours (h)	日照率 Sunshine rate (%)	平均风速 Average wind speed (m/s)
1	4.2	58.8	78	51.1	136.1	42	3.0
2	5.4	79.1	81	50.8	117.3	37	3.0
3	9.2	97.9	82	80.6	137.2	37	3.1
4	14.9	116.5	82	110.7	150.5	39	3.2
5	19.5	153.4	82	129.7	154.4	37	2.9
6	23.7	190.8	86	132.2	166.6	40	2.6
7	28.1	129.3	83	204.5	265.1	62	3.1
8	27.7	142.9	82	196.2	267.7	66	3.0
9	23.8	207.6	85	130.9	189.4	51	2.6
10	18.1	84.7	82	104.9	180.2	51	2.7
11	12.6	59.9	80	73.7	151.2	48	2.6
12	7.0	53.9	79	54.0	142.3	45	2.7
平均或合计 Mean or total	16.2	1374.2	82	1320.1	2057.9	47	2.9

1.2 主要植被类型

天童国家森林公园的山体不高，植被垂直分布并不明显，现存的都是次生植被。常绿阔叶林是这里的地带性植被和主要植被类型，组成天童常绿阔叶林的维管植物有262种，隶属于78科，其中参与构成林木层的有92种，主要以壳斗科（Fagaceae）、山茶科（Theaceae）、樟科（Lauraceae）、山矾科（Symplocaceae）和冬青科（Aquifoliaceae）的物种为主，建群种主要包括木荷（*Schima superba*）、栲树（*Castanopsis fargesii*）、米槠（*C. carlesii*）、港柯（*Lithocarpus harlandii*）和云山青冈（*Cyclobalanopsis sessilifolia*）等（图1-2-A）；在森林公园的沟谷和海拔较高的土壤瘠薄地段，常绿阔叶林中的落叶成分增加，形成外貌明显的常绿阔叶落叶阔叶混交林，建群种包括南酸枣（*Choerospondias axillaris*）、胡桃楸（*Juglans mandshurica*）、雷公鹅耳枥（*Carpinus viminea*）、檫木（*Sassafras tzumu*）、青钱柳（*Cyclocarya paliurus*）、红毒茴（*Illicium lanceolatum*）、薄叶润楠（*Machilus leptophylla*）等（图1-2-B）；公园中低海拔的山麓地带生长着大片毛竹林（图1-2-C）；在海拔590-640 m的山脊上，沿着防火线还分布有杜鹃（*Rhododendron simsii*）、白栎（*Quercus fabri*）、山矾（*Symplocos sumuntia*）、盐肤木（*Rhus chinensis*）、箬竹（*Indocalamus tessellatus*）占优势的次生灌丛（图1-2-D）[1]。

1.2 Main Vegetation Types

The existing vegetation in Tiantong National Forest Park belongs to the secondary forest. The divergence of vegetation composition are not obvious along the altitude gradient due to the small elevation differences of mountains within the park. Evergreen broad-leaved forest is the zonal and main vegetation of the park. There are 262 vascular plant species in evergreen broad-leaved forest in Tiantong belonging to 78 families. 92 of them are canopy species, dominated by Fagaceae, Theaceae, Lauraceae, Symplocaceae and Aquifoliaceae plants. *Schima superba, Castanopsis fargesii, Castanopsis carlesii, Lithocarpus harlandii, Cyclobalanopsis sessilifolia*, etc. are the major constructive species (Fig. 1-2-A). Deciduous component increases in valleys and high altitude areas with barren soil, forming the physiognomy of evergreen deciduous broad-leaved mixed forest, in which constructive species include *Choerospondias axillaris, Juglans mandshurica, Carpinus viminea, Sassafras tzumu, Cyclocarya paliurus, Illicium lanceolatum, Machilus leptophylla*, etc. (Fig. 1-2-B). There are a large area of bamboo (*Phyllostachys pubescens*) growing in the foothill at low altitude region of the park (Fig. 1-2-C). The secondary shrub distribute on the ridges at 590-640 m asl along the firebreak, dominated by *Rhododendron sissii, Quercus fabri, Symplocos sumuntia, Rhus chinensis* and *Indocalamus tessellatus* (Fig. 1-2-D) [1].

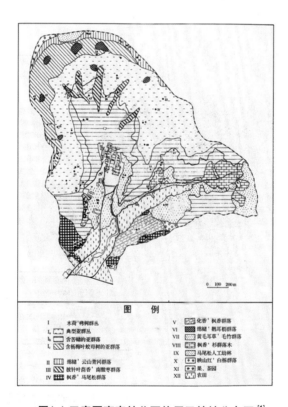

图1-1 天童国家森林公园位置及植被分布图 [1]

Fig. 1-1 Location of Tiantong National Forest Park and the vegetation map [1]

A 常绿阔叶林 Evergreen broad-leaved forest

B 常绿落叶阔叶混交林 Evergreen deciduous broad-leaved mixed forest

C 毛竹林 Bamboo (*Phyllostachys pubescens*) forest

D 灌丛 Shrub forest

图1-2 天童国家森林公园不同的植被类型
Fig. 1-2 Different types of vegetation in Tiantong National Forest Park

参考文献

1. 宋永昌，王祥荣. 1995. 浙江天童国家森林公园的植被和区系[M]. 上海：上海科学技术文献出版社.

Reference

1. Song Y C, Wang X R. 1995. Vegetation and Flora of Tiantong National Forest Park in Zhejiang province[M]. Shanghai：Shanghai Scientific and Technological Literature Press.

天童亚热带常绿阔叶林20 hm² 森林样地
The 20 hm² Subtropical Broad-leaved Evergreen Forest Plot in Tiantong

2

注：本章内容基于2010年初次调查数据进行分析，已根据2015年样地复查数据进行修正。

Note: The information of this chapter are based on first census data at 2010. The data also are corrected with second census at 2015.

2.1 样地概况

天童森林动态样地位于浙江省宁波市天童国家森林公园的核心保护区，北纬29°48.7′，东经121°47.1′。样地为长方形，东西长500 m，南北宽400 m，面积20 hm²。由于毗邻天童寺，该地常绿阔叶林一直以来作为风水林保存，长期未受较大的人为干扰，群落发育成熟，结构完整，地带性物种成分占绝对优势。仅在沟谷等不稳定微地形上，由于台风导致的滑坡等影响，间或有林窗分布，多分布有落叶植物[1]。

2.1 Overview of the Plot

The Tiantong Forest Dynamics Plot (500 m × 400 m) was established in the core zone of Tiantong National Forest Park in Ningbo, Zhejiang Province (29°48.7′N, 121°47.1′E). The forest community of this area is a mature evergreen broad-leaved forest because it neighbors Tiantong Temple and thus is protected as Fengshui forest without human disturbance for a long time. Structure of the forest is complete, and the zonal species predominate in it. Deciduous species frequently dominate in the forest gaps scattered in the valley and other unstable micro-topography, which is probably due to the landslides induced by typhoon disturbance [1].

2.2 样地建设和植被调查

2008年7月开始样地建设，2010年8月完成第一次群落调查。样地建设和群落调查参照CTFS-ForestGEO (Center for Tropical Forest Science - Forest Global Earth Observatory) 的技术规范[2]，首先，用全站仪将整个样地划分成500个20 m×20 m的样方，然后，将每个20 m×20 m样方分隔为16个5 m×5 m的样格，根据样地环境依"N"字或"Z"字形顺序调查。调查内容包括每株木本植物(胸径≥1 cm)的种类、胸径、位置、生长状况等，并悬挂铝制标牌进行个体标记[1]。

2.2 Plot Establishment and Vegetation Sampling

Tiantong plot was began to establish in July 2008, and the first vegetation census was finished in August 2010. Following the protocols from the CTFS-Forest GEO Network [2], the plot was divided into 500 quadrats of 400 m² (20 m × 20 m), each of which were further split into 16 subquadrats of 25 m² (5 m × 5 m). All stems of free-standing trees and shrubs (diameter at breast height, DBH ≥ 1 cm) were measured, tagged, mapped and identified to species [1].

2.3 地形与土壤

样地地形复杂(图2-1)，最高海拔602.9 m，最低海拔304.3 m，海拔落差298.6 m[1]。凸度介于-5.8 m到6.9 m，坡度介于13.8°到50.3°[3]。土壤主要为山地黄红壤，土壤pH值介于4.4到5.1之间，为弱酸性土壤[4]。

2.3 Topography and Soil

The topography of Tiantong plot is complicated (Fig.2-1). The maximum altitude difference is approximately 300 m (range: 304.3~602.9 m). The convexity and slope ranges from -5.8 m to 6.9 m and 13.8° to 50.3° respectively [3]. The soils are mainly red and yellow ultisols, with a pH value from 4.4 to 5.1 [4].

2.4 物种组成与群落结构

样地内共有胸径≥1 cm的独立木本植物个体94,616株，共计154种，隶属52科97属。其中，常绿树种74种，落叶树种80种。优势科依次是山茶科、樟科和壳斗科。属水平上热带区系占总属数的49.5%，

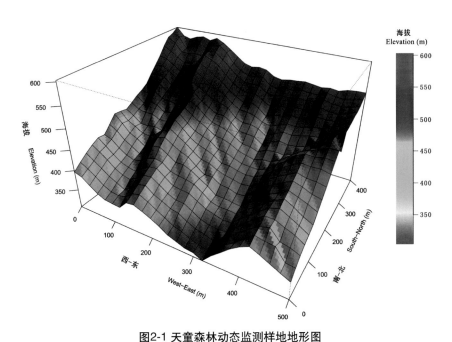

图2-1 天童森林动态监测样地地形图
Fig. 2-1 The topography map of Tiantong forest dynamics plot

温带区系占44.3%，中国特有占5.2%，世界分布占1.0%。常绿物种在样地内占绝对优势，占总重要值的80.0%。重要值最大的三个种依次是细枝柃 (*Eurya loquaiana*)、黄丹木姜子(*Litsea elongata*) 和南酸枣。稀有种共计57种，占总物种数的37.0%，其中落叶种类较多，共有37种，占稀有种总数的64.9%。稀有种的多度仅占总多度的0.2%，占总重要值的0.4%。种–面积曲线符合对数分布模型 (图2-3)，模型拟合方程为：$S=18.54\ln(A)+100.11$ ($R^2=0.9921$, $P<0.001$)，其中，S为物种数，A为样方面积。样地内个体的平均胸径5.66 cm，胸径最大的个体是南酸枣 (DBH=87.5 cm)。木本植物整体径级结构呈逆"J"字型 (图2-2)，胸径4 cm以下的个体占总个体数的64.15%。

样地群落垂直结构清晰，分为乔木层、亚乔木层、灌木层。乔木层以南酸枣、港柯、云山青冈、木荷、栲树为优势种，亚乔木层以黄丹木姜子、杨梅叶蚊母树 (*Distylium myricoides*)、浙江新木姜子 (*Neolitsea aurata* var. *chekiangensis*)、红淡比 (*Cleyera japonica*)、薄叶山矾 (*Symplocos anomala*) 为优势种，灌木层以细枝柃、毛花连蕊茶 (*Camellia fraterna*) 等为优势种。

2.4 Species composition and Community structure

In total, 94,616 independent woody individuals, belonging to 154 species (74 evergreen species and 80 deciduous species), 97 genera and 52 families, were recorded. The first three dominant families were Theaceae, Lauraceae and Fagaceae. Regarding the floristic composition in Tiantong plot, at the genera level, the tropical element has the hightest proportion (49.5% in total). The temperate element comes next, accounting for 44.3%. The endemic element of China accounts for 5.2%. The percentage of the cosmopolitan element is very low (1.0% in total). Evergreen species dominate in the plot, and the important value (IV) is 80.0. The top three dominant species according to their IVs were *Eurya loquaiana*, *Litsea elongate* and *Choerospondias axillaris*. There were 57 rare species, 64.9% of which (37/57) were deciduous species. The abundance and IV of rare species make up 0.2% and 0.4% of total abundance and total IV. The species-area curve of Tiantong plot was best fitted by the logarithmic distribution model ($S = 18.54*\ln(A) + 100.11$, $R^2=0.9921$, $P<0.001$) (Fig.2-3) where S is the number of species, A is the plot area. The mean DBH of individuals in the plot is 5.66 cm. The individual having the largest DBH is *Choerospondias axillaris* (DBH=87.5 cm). The diameter distribution of the forest followed a reversal J-shape (Fig. 2-2). Individuals with DBH < 4 cm account for 64.15% of the total number of individuals in the forest.

The vertical structure of community in Tiantong plot could be distinctly divided into three layers–canopy layer, subcanopy layer and shrub layer. *Choerospondias axillaris, Lithocarpus harlandii, Cyclobalanopsis sessilifolia, Schima superba* and *Castanopsis fargesii* were dominant species in canopy layer; *Litsea elongate, Distylium myricoides, Neolitsea aurata* var. *chekiangensis, Cleyera japonica* and *Symplocos anomala* predominated in subcanopy layer; *E. loquaiana* and *Camellia fraternal* prevailed in shrub layer.

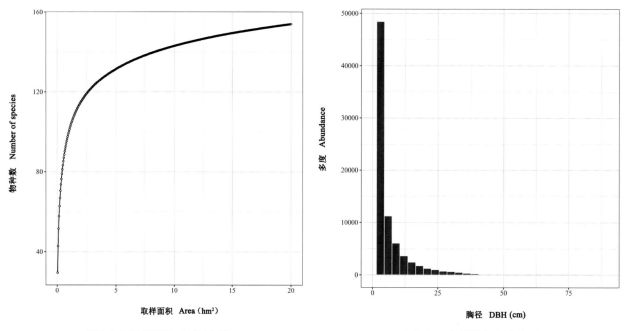

图2-2 天童样地种–面积曲线
Fig.2-2 Species-area curve of Tiantong plot

图2-3 天童样地径级分布图
Fig.2-3 Size class of DBH distribution of Tiantong plot

参考文献

1. 杨庆松, 马遵平, 谢玉彬, 等. 2011. 浙江天童 20 hm² 常绿阔叶林动态监测样地的群落特征[J]. 生物多样性, 19:215-223.
2. Condit R. 1998. Tropical forest census plots: Methods and results from Barro Colorado Island, Panama and a comparison with other plots[J]. Springer, Berlin.
3. Yang Q S, Shen G C, Liu H M, et al. 2016. Detangling the effects of environmental filtering and dispersal limitation on aggregated distributions of tree and shrub species: Life stage matters[J]. PLoS ONE 11:e0156326.
4. 张娜, 王希华, 郑泽梅, 等. 2012. 浙江天童常绿阔叶林土壤的空间异质性及其与地形的关系[J]. 应用生态学报, 23:2361-2369.

Reference

1. Yang Q S, Ma Z P, Xie Y B, et al. 2011. Community structure and species composition of an evergreen broad-leaved forest in Tiantong's 20 ha dynamic plot, Zhejiang Province, Eastern China[J]. Biodiversity Science, 19:215-223.
2. Condit, R. 1998. Tropical forest census plots: methods and results from Barro Colorado Island, Panama and a comparison with other plots. Springer, Berlin.
3. Yang Q S, Shen G C, Liu H M, et al. 2016. Detangling the effects of environmental fifiltering and dispersal limitation on aggregated distributions of tree and shrub species: Life stage matters[J]. PLoS ONE 11:e0156326.
4. Zhang N, Wang X H, Zheng Z M, et al. 2012. Spatial heterogeneity of soil properties and its relationships with terrain factors in broad-leaved forest in Tiantong of Zhejiang Province, East China[J]. Chinese Journal of Applied Ecology 23:2361-2369.

树种及其空间分布
Tree Species and their Spatial Distribution 3

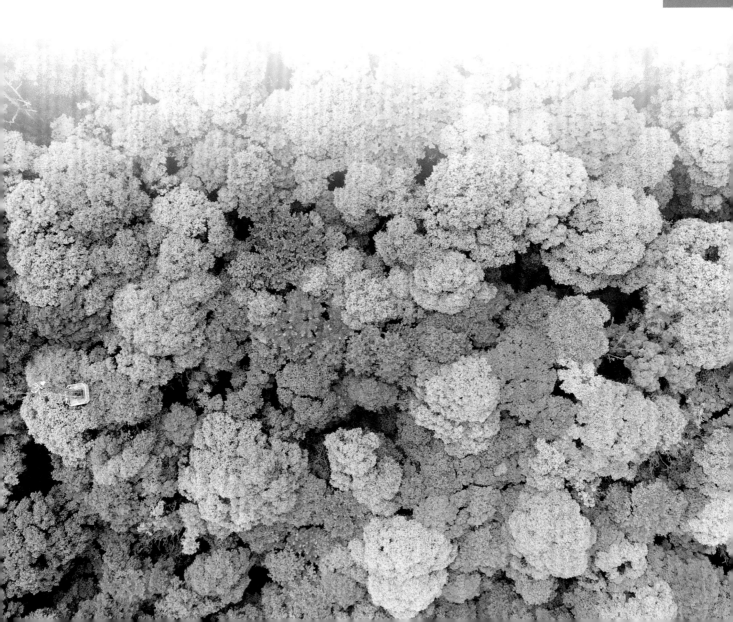

1 银杏

Yín xìng | Maidenhairtree

Ginkgo biloba Linn.
银杏科 | Ginkgoaceae

代码 (SpCode) = GINBIL
个体数 (Individual number/20 hm^2) = 5
最大胸径 (Max DBH) = 11.5 cm
重要值排序 (Importance value rank) = 123

落叶乔木，高达40 m。树皮灰褐色，纵裂。枝有长枝、短枝之分。叶扇形，上部宽5~8 cm，上缘有浅或深的波状缺刻，有时中部缺裂较深，基部楔形，有长柄；叶在短枝上簇生。种子核果状，近球形，具长梗，成熟时黄或橙黄色，被白粉。花期3~4月，果期9~10月。

Deciduous trees, up to 40 m tall; bark gray or grayish brown, longitudinally fissured; two kinds of branchlets, long and short. Leaf sector, with wave crack and long petiole. Seeds elliptic, narrowly obovoid or subglobose, sarcotesta yellow, or orange-yellow glaucous when ripe. Fl. Mar. - Apr., fr. Sep. - Oct..

树干　　Trunk
摄影：杨庆松　Photo by: Yang Qingsong

叶　　Leaves
摄影：杨庆松　Photo by: Yang Qingsong

花枝　　Flowering branch
摄影：杨庆松　Photo by: Yang Qingsong

径级分布表 DBH class

胸径区间 (Diameter class) (cm)	个体数 (No. of individuals in the plot)	比例 (Proportion) (%)
1~2	3	60.00
2~5	1	20.00
5~10	0	0.00
10~20	1	20.00
20~30	0	0.00
30~60	0	0.00
≥60	0	0.00

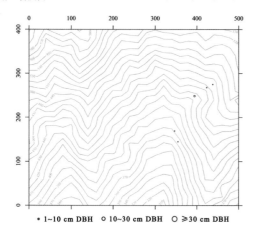

• 1~10 cm DBH　○ 10~30 cm DBH　◯ ≥30 cm DBH

个体分布图 Distribution of individuals

2 杉木

Shān mù | China fir

Cunninghamia lanceolata (Lamb.) Hook.
杉科 | Taxodiaceae

代码 (SpCode) = CUNLAN
个体数 (Individual number/20 hm^2) = 17
最大胸径 (Max DBH) = 22.1 cm
重要值排序 (Importance value rank) = 100

常绿乔木，高达30 m。树皮深灰色至暗褐色或红棕色，不规则条状纵裂；树冠圆锥形。叶在主枝上辐射伸展，在侧枝基部扭转成2列，革质，坚硬，披针形，扁平，有细锯齿，长3~6 cm。球果近球形或圆卵形。花期4月；球果10月成熟。

Evergreen trees, up to 30 m tall; bark dark gray to dark brown, or reddish brown, longitudinally fissured, cracking into irregular flakes; crown pyramidal. Leaves glossy deep green adaxially, narrowly linear-lanceolate, 3~6 cm, margin denticulate. Seed ovoid or subglobose. Fl. Apr., fr. Oct..

植株　　Whole plant
摄影：杨庆松　Photo by: Yang Qingsong

枝叶　　Branchs and leaves
摄影：杨庆松　Photo by: Yang Qingsong

果枝　　Fruiting branch
摄影：杨庆松　Photo by: Yang Qingsong

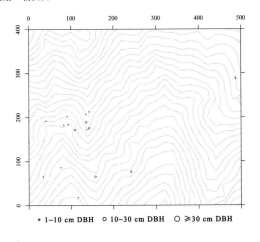

个体分布图　Distribution of individuals

径级分布表 DBH class

胸径等级 (Diameter class) (cm)	个体数 (No. of individuals in the plot)	比例 (Proportion) (%)
1~2	6	35.29
2~5	2	11.76
5~10	4	23.53
10~20	4	23.53
20~30	1	5.88
30~60	0	0.00
≥60	0	0.00

3 三尖杉

Sān jiān shān | Fortune Plumyew

Cephalotaxus fortunei Hook.
三尖杉科 | Cephalotaxaceae

代码 (SpCode) = CEPFOR

个体数 (Individual number/20 hm^2) = 33

最大胸径 (Max DBH) = 8.4 cm

重要值排序 (Importance value rank) = 99

常绿乔木，高达20 m。树皮褐色或红褐色，裂成片状脱落。叶片深绿色，正面光滑，披针状线形，通常微弯，下面气孔带约1~2 mm，边缘宽度0.1~0.4 mm。假种皮初时黄色或绿色，成熟时变紫。种子椭圆形。花期4~5月，果期6~10月。

Evergreen trees, 20 m tall; bark dark reddish brown, peeling in strips. Leaves blade deep green and glossy adaxially, linear-lanceolate, falcate, stomatal bands 1~2 mm wide, marginal bands 0.1~0.4 mm wide. Aril yellow or green initially, turning purple when ripe. Seeds ellipsoid. Fl. Apr. - May, fr. Jun. - Oct..

幼苗　Seedling
摄影：杨庆松　Photo by: Yang Qingsong

枝叶　Branch and leaves
摄影：杨庆松　Photo by: Yang Qingsong

果　Fruits
摄影：严靖　Photo by: Yan Jing

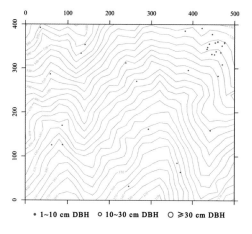

个体分布图　Distribution of individuals

径级分布表　DBH class

胸径区间 (Diameter class) (cm)	个体数 (No. of individuals in the plot)	比例 (Proportion) (%)
1~2	13	39.39
2~5	12	36.36
5~10	8	24.24
10~20	0	0.00
20~30	0	0.00
30~60	0	0.00
≥60	0	0.00

4 榧树

Fěi shù | China Torreya

Torreya grandis Fort. ex Lindl.
红豆杉科 | Taxaceae

代码 (SpCode) = TORGRA
个体数 (Individual number/20 hm^2) = 1
最大胸径 (Max DBH) = 38.6 cm
重要值排序 (Importance value rank) = 110

常绿乔木，高达25 m。树皮不规则纵裂，淡黄灰色、深灰色或灰褐色。叶线形，表面深绿色，光亮，背面气孔带宽(0.2) 0.3～0.4 mm。种子熟时假种皮淡紫褐色，有白粉。种子椭圆形到卵形，长椭圆形，倒卵形，或倒卵球形圆锥形，2～4.5×1.2～2.5 cm。花期4月，果期翌年9～11月。

Evergreen trees, up to 25 m tall; bark with irregular vertical fissures, light yellowish gray, dark gray, or grayish brown. Leaves bright green and glossy adaxially, linear-lanceolate, blade bright green and glossy adaxially, stomatal bands (0.2-) 0.3~0.4 mm wide. Aril pale purplish brown and white powdery when ripe. Seed ellipsoid to ovoid, elongate-ellipsoid, obovoid, or obovoid-conical, 2~4.5 × 1.2~2.5 cm. Fl. Apr., fr. Sep. - Nov. of following year.

树干　　Trunk
摄影：杨庆松　Photo by: Yang Qingsong

叶　　Leaves
摄影：杨庆松　Photo by: Yang Qingsong

果枝　　Fruiting branch
摄影：汪远　Photo by: Wang Yuan

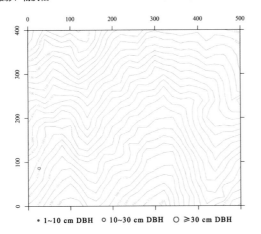

个体分布图 Distribution of individuals

径级分布表 DBH class

胸径等级 (Diameter class) (cm)	个体数 (No. of individuals in the plot)	比例 (Proportion) (%)
1～2	0	0.00
2～5	0	0.00
5～10	0	0.00
10～20	0	0.00
20～30	0	0.00
30～60	1	100.00
≥60	0	0.00

5 响叶杨

Xiǎng yè yáng | China Aspen

Populus adenopoda Maxim.
杨柳科 | Salicaceae

代码 (SpCode) = POPADE
个体数 (Individual number/20 hm^2) = 2
最大胸径 (Max DBH) = 45.7 cm
重要值排序 (Importance value rank) = 96

落叶乔木，高15～30 m。树皮灰白色，光滑，老时深灰色，纵裂。叶柄侧扁，顶端有2显著腺点。叶卵状圆形或卵形，长5～15 cm，宽4～7 cm，边缘有内曲圆锯齿，齿端有腺点；蒴果卵状长椭圆形，(2) 4～6 mm, 无毛。种子暗褐色。花期3～4月，果期4～5月。

Decdious trees, up to 15-30 m tall; bark grayish white, smooth, becoming dark gray, furrowed. Petiole very laterally flattened, with 2 raised glands; leaf blade ovate-orbicular or ovate, 5～15 × 4～7(～13) cm, base truncate or cordate, rarely subrounded or cuneate, margin incurved, glandular crenate-serrate or loosely or coarsely dentate. Capsule long ovoid-ellipsoid, (2～) 4～6 mm, glabrous. Seeds dark. Fl. Mar. - Apr., fr. Apr. - May.

树干　　　　Trunk
摄影：杨庆松　Photo by: Yang Qingsong

果枝　　　　Fruiting branch
摄影：汪远　　Photo by: Wang Yuan

幼叶　　　　Young leaves
摄影：杨庆松　Photo by: Yang Qingsong

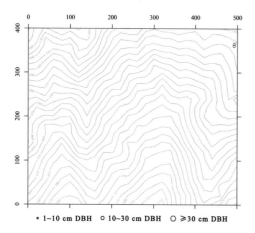

个体分布图 Distribution of individuals

径级分布表 DBH class

胸径区间 (Diameter class) (cm)	个体数 (No. of individuals in the plot)	比例 (Proportion) (%)
1～2	0	0.00
2～5	0	0.00
5～10	0	0.00
10～20	0	0.00
20～30	0	0.00
30～60	2	100.00
≥60	0	0.00

6 杨梅 | Yáng méi | China Bayberry

Myrica rubra (Lour.) Sieb. et Zucc.
杨梅科 | Myricaceae

代码 (SpCode) = MYRRUB
个体数 (Individual number/20 hm^2) = 79
最大胸径 (Max DBH) = 28.7 cm
重要值排序 (Importance value rank) = 62

常绿乔木，高达15 m，树皮灰色。叶片基部楔形倒卵形或椭圆状倒卵形，叶革质，无毛，稀中上部疏生锐齿，常密生小枝上部。花雌雄异株。核果球形，熟时深红或紫红色，球状，直径1~1.5 cm，栽培品种可达3 cm，具乳头状凸起，果皮肉质，可食用，味酸甜。花期4月，果期6~7月。

Evergreen trees, to 15 m tall; bark gray. Leaf blade cuneate-obovate or narrowly elliptic-obovate, leathery, glabrous, margin entire or serrate in apical 1/2, apex obtuse to acute; densely grow on the top branch. Dioecism. Drupe dark red or purple-red at maturity, globose, 1~1.5 cm in diam., to 3 cm when cultivated, papilliferous; pericarp edible, taste sour and sweet. Fl. Apr., fr. Jun. - Jul..

树干 Trunk
摄影：杨庆松 Photo by: Yang Qingsong

叶 Leaves
摄影：杨庆松 Photo by: Yang Qingsong

花枝 Flowering branch
摄影：杨庆松 Photo by: Yang Qingsong

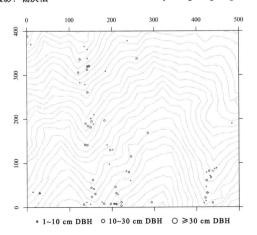

个体分布图 Distribution of individuals

径级分布表 DBH class

胸径等级 (Diameter class) (cm)	个体数 (No. of individuals in the plot)	比例 (Proportion) (%)
1~2	7	8.86
2~5	7	8.86
5~10	23	29.11
10~20	35	44.30
20~30	7	8.86
30~60	0	0.00
≥60	0	0.00

7 青钱柳　　　　　　　　　　　　　　　Qīng qián liǔ | Cyclocarya

Cyclocarya paliurus (Batal.) Iljinsk.
胡桃科 | Juglandaceae

代码 (SpCode) = CYCPAL
个体数 (Individual number/20 hm^2) = 60
最大胸径 (Max DBH) = 54.5 cm
重要值排序 (Importance value rank) = 51

落叶乔木，高达30 m。枝条髓部薄片状分隔。奇数羽状复叶长，具(5~)7~9(~11)小叶；小叶长椭圆状卵形或宽披针形，基部歪斜，宽楔形或近圆，先端钝或锐尖；侧生小叶及无柄或小叶柄2 mm，顶生小叶叶柄1~15 mm；雌雄同株；小坚果扁球状，长约7 mm；盘状翅革质，圆形至卵形，2.5~6 cm。花期4~5月，果期7~9月。

Deciduous trees, up to 30 m tall. Branch pith lamelliform separation; imparipinnate leaves; leaflets (5) 7~9(11); leaflets blade elliptic-ovate to broadly lanceolate, base oblique, broadly cuneate to subrounded, apex obtuse or acute; lateral leaflets sessile or petiolule to 2 mm, terminal petiolule 1~15 mm.; monoecism; Nutlets compressed globose, 7 mm; disc wing leathery, orbicular to ovate, 2.5~6 cm. Fl. Apr. - May, fr. Jul. - Sep..

树干　　Trunk
摄影：杨庆松　　Photo by: Yang Qingsong

叶　　Leaves
摄影：杨庆松　　Photo by: Yang Qingsong

花枝　　Flowering branch
摄影：杨庆松　　Photo by: Yang Qingsong

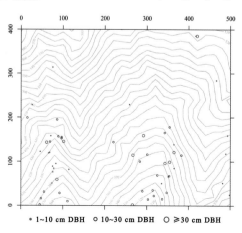
个体分布图 Distribution of individuals

径级分布表 DBH class

胸径区间 (Diameter class) (cm)	个体数 (No. of individuals in the plot)	比例 (Proportion) (%)
1~2	3	5.00
2~5	16	26.67
5~10	6	10.00
10~20	17	28.33
20~30	8	13.33
30~60	10	16.67
≥60	0	0.00

8 胡桃楸 (华东野核桃)

Hú táo qiū | Manchurian Walnut

Juglans mandshurica Maxim.
胡桃科 | Juglandaceae

代码 (SpCode) = JUGMAN
个体数 (Individual number/20 hm^2) = 55
最大胸径 (Max DBH) = 39.2 cm
重要值排序 (Importance value rank) = 54

落叶乔木，可达25 m高。叶40～90 cm；叶柄5～23 cm；叶轴疏生腺状短柔毛，有时密；小叶 (7) 9～19，侧生小叶无柄，基部偏斜，近心形，边缘有锯齿，很少有细锯齿，先端渐尖；顶生小叶柄1～5 cm。坚果卵球形或椭圆形，3～7.5×3～5 cm。花期4～5月，果期8～10月。

Deciduous trees, to 25 m tall. Leaves 40~90 cm; petiole 5~23 cm; petiole and rachis sparsely to moderately glandular pubescent, occasionally densely; leaflets (7) 9~19, lateral ones sessile, base oblique, subcordate, margin serrate, rarely serrulate, apex acuminate; terminal petiolule 1~5 cm. Nuts globose, ovoid, or ellipsoid, 3~7.5 × 3~5 cm. Fl. Apr. - May, fr. Aug. - Oct..

果　　Fruits
摄影：杨庆松　Photo by: Yang Qingsong

叶　　Leaves
摄影：杨庆松　Photo by: Yang Qingsong

花枝　　Flowering branch
摄影：杨庆松　Photo by: Yang Qingsong

个体分布图 Distribution of individuals

径级分布表 DBH class

胸径等级 (Diameter class) (cm)	个体数 (No. of individuals in the plot)	比例 (Proportion) (%)
1～2	0	0.00
2～5	4	7.27
5～10	1	1.82
10～20	26	47.27
20～30	20	36.36
30～60	4	7.27
≥60	0	0.00

9 化香树   Huà xiāng shù | Dyetree

Platycarya strobilacea Sieb. et Zucc.
胡桃科 | Juglandaceae

代码 (SpCode) = PLASTR
个体数 (Individual number/20 hm^2) = 25
最大胸径 (Max DBH) = 31.4 cm
重要值排序 (Importance value rank) = 88

落叶乔木，高达15 m。奇数羽状复叶，具 (3) 7~23小叶；叶柄1.2~9.2 cm；小叶纸质，对生，具重锯齿，侧生小叶无柄，叶片卵状披针形至狭椭圆状披针形，基部歪斜；顶生小叶，基部圆形或宽楔形。雌雄同株，穗状花序2~10 cm。果序球果状，常宿存。种子卵圆形。花期5~6月，果期7~10月。

Decdious trees, up to 15 m tall. Imparipinnate compound leaves, leaflets (3~) 7~23 cm; petiole 1.2~9.2 cm; leaflets papery; opposite, double serrate; lateral leaflets sessile, blade ovate-lanceolate to narrowly elliptic-lanceolate, base oblique to cuneate; terminal leaflet with petiolule, base rounded or broadly cuneate. Androgynous spike 2~10 cm. Fruiting subglobose; persistent. Nutlets suborbicular to obovate. Fl. May - Jun., fr. Jul. - Oct..

树干   Trunk
摄影：杨庆松   Photo by: Yang Qingsong

叶   Leaves
摄影：杨庆松   Photo by: Yang Qingsong

花枝   Flowering branch
摄影：杨庆松   Photo by: Yang Qingsong

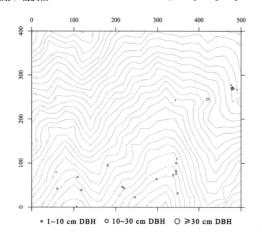
个体分布图 Distribution of individuals

径级分布表 DBH class

胸径区间 (Diameter class) (cm)	个体数 (No. of individuals in the plot)	比例 (Proportion) (%)
1~2	0	0.00
2~5	3	12.00
5~10	4	16.00
10~20	15	60.00
20~30	2	8.00
30~60	1	4.00
≥60	0	0.00

10 雷公鹅耳枥

Léi gōng é ěr lì | Thunder god Hornbeam

Carpinus viminea Lindl.
桦木科 | Betulaceae

代码 (SpCode) = CARVIM
个体数 (Individual number/20 hm^2) = 874
最大胸径 (Max DBH) = 61.3 cm
重要值排序 (Importance value rank) = 14

落叶乔木，高达20 m，树皮深灰色。小枝棕色，密生白色皮孔，无毛。叶柄长1～3 cm，无毛。叶宽椭圆形、长圆形或卵状披针形，背面疏生长柔毛沿中脉和侧脉，基部心形，很少圆楔形或近圆形，边缘有重锯齿，先端尖、渐尖或尾状，基部近心形。小坚果宽卵球形，长3～4 mm，宽 2.5～3.5 mm，顶端被长柔毛，上部疏被树脂腺体，具纵肋。花期3～4月，果期9月。

Decdious trees, up to 20 m tall; bark dark gray. Branchlets brown, glabrous. Petiole 1~3 cm, glabrous; leaf blade elliptic, oblong, or ovate-lanceolate, abaxially sparsely villous along midvein and lateral veins, base subcordate, rarely rounded-cuneate or subrounded, margin doubly mucronate serrate, apex acuminate, acute, or caudate. Nutlet broadly ovoid, 3~4 × 2.5~3.5 mm, glabrous except villous at apex, sparsely resinous glandular, prominently ribbed. Fl. Mar. - Apr., fr. Sep..

树干　　Trunk
摄影：杨庆松　　Photo by: Yang Qingsong

枝叶　　Branch and leaves
摄影：杨庆松　　Photo by: Yang Qingsong

果枝　　Fruiting branch
摄影：王樟华　　Photo by: Wang Zhanghua

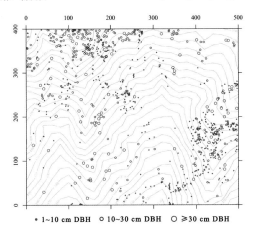

个体分布图　Distribution of individuals

径级分布表 DBH class

胸径等级 (Diameter class) (cm)	个体数 (No. of individuals in the plot)	比例 (Proportion) (%)
1～2	22	2.52
2～5	215	24.60
5～10	306	35.01
10～20	133	15.22
20～30	108	12.36
30～60	89	10.18
≥60	1	0.11

11 米槠   Mǐ zhū | Carles Oatchesnut

Castanopsis carlesii (Hemsl.) Hayata
壳斗科 | Fagaceae

代码 (SpCode) = CASCAR
个体数 (Individual number/20 hm^2) = 537
最大胸径 (Max DBH) = 64.6 cm
重要值排序 (Importance value rank) = 16

常绿乔木，高达20 m。小枝及花序轴被微被红褐色片状蜡鳞。叶披针形或卵状披针形，革质，先端渐尖或稍尾尖，全缘或中部以上具浅齿，幼叶下面被红褐或褐黄色蜡鳞层，老叶稍灰白色。坚果近球形或宽圆锥形，先端短尖。花期4～6月，果期翌年9～11月。

Evergreen trees, up to 20 m tall. Young shoots and rachis of inflorescences sparsely covered with reddish brown, lamellate, waxy scalelike trichomes. Leaf blade lanceolate to ovate, leathery, apex acuminate to narrowly caudate, margin entire or with a few shallow teeth, abaxially with layers of reddish brown to yellowish brown, lamellate scalelike trichomes when young but grayish brown to silvery with age. Nut subglobose to broadly conical, apex shortly pointed. Fl. Apr. - Jun., fr. Sep. - Nov. of following year.

树干   Trunk
摄影：杨庆松   Photo by: Yang Qingsong

枝叶   Branch and leaves
摄影：杨庆松   Photo by: Yang Qingsong

花枝   Flowering branches
摄影：杨庆松   Photo by: Yang Qingsong

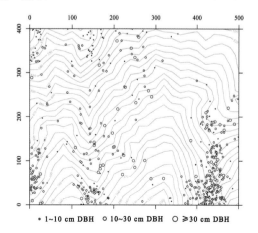

个体分布图 Distribution of individuals

径级分布表 DBH class

胸径区间 (Diameter class) (cm)	个体数 (No. of individuals in the plot)	比例 (Proportion) (%)
1～2	83	15.46
2～5	70	13.04
5～10	35	6.52
10～20	134	24.95
20～30	148	27.56
30～60	65	12.10
≥60	2	0.37

12 栲树　　　　　　　　　　　　　　　　　　　Kǎo shù | Farges's Oatchestnut

Castanopsis fargesii Franch.
壳斗科 | Fagaceae

代码 (SpCode) = CASFAR
个体数 (Individual number/20 hm^2) = 746
最大胸径 (Max DBH) = 77.8 cm
重要值排序 (Importance value rank) = 9

常绿乔木，高达30 m。芽鳞、幼枝上半部及叶下面均被易脱落的红褐或灰褐色蜡鳞层，枝、叶无毛。叶柄长1～2 cm。叶长7～15 cm，基部圆或宽楔形，全缘或近顶部疏生浅齿，先端短尖或渐尖。壳斗球形或宽卵圆形，连刺径2.5～3 cm；坚果圆锥形，无毛。花期4～5月，果期翌年8～10月。

Evergreen trees, up to 30 m tall. Bud scales, young branchlets from middle to apex, and leaf blades abaxially covered with glabrescent, rust-colored, small, lamellate, waxy scalelike trichomes, branches and leaves glabrous. Petiole 1~2 cm; leaf blade 7~15 cm, base rounded to broadly cuneate and sometimes inaequilateral, margin entire or sometimes with few shallow teeth from middle to apex, apex acute to acuminate. Cupule globose to broadly ovoid, 2.5~3 cm in diam. Nut conical, glabrous. Fl. Apr. - May, fr. Aug. - Oct. of following year.

幼苗　Seedling
摄影：刘何铭　Photo by: Liu Heming

叶背　Leaf abaxial surface
摄影：杨庆松　Photo by: Yang Qingsong

花枝　Flowering branch
摄影：杨庆松　Photo by: Yang Qingsong

径级分布表　DBH class

胸径等级 (Diameter class) (cm)	个体数 (No. of individuals in the plot)	比例 (Proportion) (%)
1～2	55	7.37
2～5	61	8.18
5～10	39	5.23
10～20	131	17.56
20～30	246	32.98
30～60	204	27.35
≥60	10	1.34

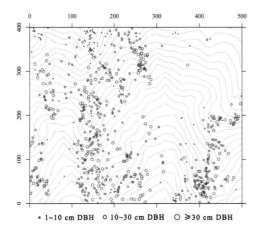

个体分布图　Distribution of individuals

13 苦槠

Kǔ zhū | Hardleaf Oatchestnut

Castanopsis sclerophylla (Lindl.) Schott.
壳斗科 | Fagaceae

代码 (SpCode) = CASSCL

个体数 (Individual number/20 hm^2) = 23

最大胸径 (Max DBH) = 49.2 cm

重要值排序 (Importance value rank) = 87

常绿乔木，高达15 m。枝、叶无毛。叶柄长1.5～2.5 cm，叶长椭圆形、卵状椭圆形或倒卵状椭圆形，长7～15 cm，革质，老叶下面银灰色，基部宽楔形或近圆，中部以上具锯齿，稀全缘，先端短尖或短尾状。花序无毛，雌花序长约15 cm。壳斗近球形，几全包果，径1.2～1.5 cm，不规则瓣裂，外部黄棕色。坚果近球形，可食用。花期4～5月，果期10～11月。

Evergreen trees, up to 15m; branches and leaf blades glabrous. Petiole 1.5~2.5 cm; leaf blade oblong, ovate-elliptic, or obovate-elliptic, 7~15 cm, leathery, adaxially silver-gray with age, base rounded to broadly cuneate and usually inaequilateral, margin from middle to apex serrulate or rarely entire, apex acuminate, cuspidate, or shortly caudate. Female inflorescence ca. 15 cm, glabrous. Cupule globose to subglobose, 1.2~1.5 cm in diam., completely or almost completely enclosing nut, irregularly valved, outside yellowish brown puberulent. Nut subglobose, edible. Fl. Apr. - May, fr. Oct. - Nov..

树干　　Trunk
摄影：杨庆松　Photo by: Yang Qingsong

叶　　Leaf
摄影：杨庆松　Photo by: Yang Qingsong

花枝　　Flowering branch
摄影：杨庆松　Photo by: Yang Qingsong

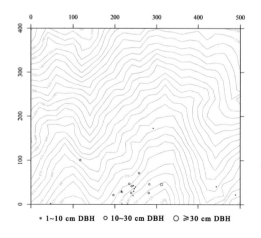

个体分布图　Distribution of individuals

径级分布表　DBH class

胸径区间 (Diameter class) (cm)	个体数 (No. of individuals in the plot)	比例 (Proportion) (%)
1～2	1	4.35
2～5	7	30.43
5～10	5	21.74
10～20	7	30.43
20～30	2	8.70
30～60	1	4.35
≥60	0	0.00

14 赤皮青冈

Chì pí qīng gāng | Redbark Oak

Cyclobalanopsis gilva (Blume) Oerst.
壳斗科 | Fagaceae

代码 (SpCode) = CYCGIL
个体数 (Individual number/20 hm^2) = 68
最大胸径 (Max DBH) = 29.9 cm
重要值排序 (Importance value rank) = 75

常绿乔木，高达30 m。小枝密被灰黄或黄褐色星状绒毛。叶倒披针形或倒卵状长椭圆形，中部以上具锯齿，老叶下面被灰黄色星状短绒毛。果序密被灰黄色绒毛。壳斗碗状，具6~7环带，疏被毛；坚果倒卵状椭圆形，顶端被微柔毛，果脐微凸。花期5月，果期10月。

Evergreen trees, up to 30 m tall; branchlets densely pale orangish brown or tawny stellate tomentose. Leaf blade oblanceolate or obovate-elliptic, margin apical 1/2 with awnlike serrations. Old leaves densely pale orangish brown stellate short tomentose. Infructescence densely pale orangish brown tomentose. Cupule bowl-shaped, with 6~7 rings, sparsely hairs. Nut obovate-elliptic, apex puberulent, slightly raised. Fl. May, fr. Oct..

树干　　Trunk
摄影：胡瑾瑾　Photo by: Hu Jinjin

枝叶　　Branch and leaves
摄影：杨庆松　Photo by: Yang Qingsong

花枝　　Flowering branches
摄影：汪远　Photo by: Wang Yuan

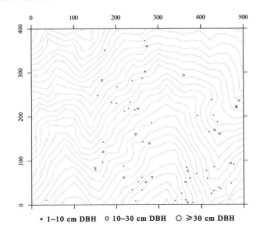

个体分布图 Distribution of individuals

径级分布表 DBH class

胸径等级 (Diameter class) (cm)	个体数 (No. of individuals in the plot)	比例 (Proportion) (%)
1~2	5	7.35
2~5	13	19.12
5~10	30	44.12
10~20	16	23.53
20~30	4	5.88
30~60	0	0.00
≥60	0	0.00

15 青冈

Qīng gāng | Blue Japanese Oak

Cyclobalanopsis glauca (Thunb.) Oerst.
壳斗科 | Fagaceae

代码 (SpCode) = CYCGLA
个体数 (Individual number/20 hm^2) = 102
最大胸径 (Max DBH) = 41.1 cm
重要值排序 (Importance value rank) = 64

常绿乔木，高达20 m。小枝无毛。叶片革质，叶倒卵状椭圆形或长椭圆形，中部以上具锯齿，上面无毛，下面常被灰白色粉霜；叶柄长1~3 cm。壳斗碗状，疏被毛，具5~6环带；坚果长卵圆形或椭圆形，近无毛。花期4~5月，果期10月。

Evergreen trees, up to 20 m tall; branchlets glabrous. Leaf obvate-elliptic or oblong-elliptic; leathery, margin apical 1/2 remotely serrate, glabrous above, base grayish white pruinous; petiole 1~3 cm. Cupule bowl-shaped, sparsely hairs, with 5~6 rings. Nut oblong-ovoid, or ellipsoid, glabrous. Fl. Apr. - May, fr. Oct..

树干　　　　　Trunks
摄影：杨庆松　　Photo by: Yang Qingsong

枝叶　　　　　Branch and leaves
摄影：杨庆松　　Photo by: Yang Qingsong

果　　　　　Fruit
摄影：杨庆松　　Photo by: Yang Qingsong

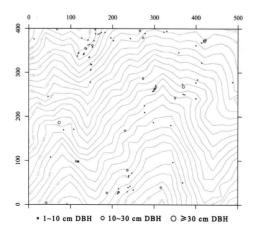

个体分布图 Distribution of individuals

径级分布表 DBH class

胸径区间 (Diameter class) (cm)	个体数 (No. of individuals in the plot)	比例 (Proportion) (%)
1~2	13	12.75
2~5	35	34.31
5~10	34	33.33
10~20	14	13.73
20~30	4	3.92
30~60	2	1.96
≥60	0	0.00

16 细叶青冈

Xì yè qīng gāng | Thinleaf Oak

Cyclobalanopsis gracilis (Rehd. et Wils.) Cheng et T. Hong
壳斗科 | Fagaceae

代码 (SpCode) = CYCGRA
个体数 (Individual number/20 hm^2) = 1084
最大胸径 (Max DBH) = 38.2 cm
重要值排序 (Importance value rank) = 24

常绿乔木，高达15 m，树皮灰褐色。叶片长卵形至卵状披针形，叶缘1/3以上有细尖锯齿，叶面亮绿色，叶背灰白色，有伏贴单毛；叶柄长1～1.5 cm。雌雄异花。壳斗碗形，包着坚果1/3～1/2，外壁被伏贴灰黄色绒毛；小苞片合生成6～9条同心环带。坚果椭圆形，果脐微凸起。花期3～4月，果期10～11月。

Evergreen trees, up to 15 m tall; bark gray or grayish brown. Leaf blade oblong-ovate to ovate-lanceolate, margin apical 1/3 sharply serrulate, adaxially bright green, abaxially grayish white and with prostrate simple hairs; Petiole 1~1.5 cm. Diclinism. Cupule bowle-shaped, enclosing 1/3~1/2 of nut, out side with prostrate grayish yellow tomentose hairs. Bracts in 6~9 concentric rings. Nut ellipsoid, slightly convex. Fl. Mar. - Apr, fr. Oct. - Nov..

树干 Trunk
摄影：杨庆松 Photo by: Yang Qingsong

种子 Seed
摄影：刘何铭 Photo by: Liu Heming

枝叶 Branch and leaves
摄影：杨庆松 Photo by: Yang Qingsong

径级分布表 DBH class

胸径等级 (Diameter class) (cm)	个体数 (No. of individuals in the plot)	比例 (Proportion) (%)
1～2	37	3.41
2～5	334	30.81
5～10	529	48.80
10～20	142	13.10
20～30	28	2.58
30～60	14	1.29
≥60	0	0.00

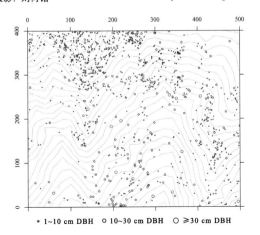

个体分布图 Distribution of individuals

17 小叶青冈 (青栲)   Xiao yè qīng gāng | Littleleaf Oak

Cyclobalanopsis myrsinifolia (Blume) Oerst.
壳斗科 | Fagaceae

代码 (SpCode) = CYCMYR
个体数 (Individual number/20 hm^2) = 83
最大胸径 (Max DBH) = 56.7 cm
重要值排序 (Importance value rank) = 50

常绿乔木，高达20 m。小枝无毛。叶卵状披针形或椭圆状披针形，中部以上具细齿，上面绿色，下面苍灰色，无毛。壳斗杯状，壁薄，被灰白色柔毛，内壁无毛，具6~9环带；坚果卵圆形或椭圆形，无毛，果脐平。花期6月，果期10月。

Evergreen trees, up to 20 m tall. Branchlets glabrous. Leaf blade ovate-lanceolate to elliptic-lanceolate, margin apical 1/2 serrulate, adaxially green, abaxially dark gray, glabrous. Cupule copular, wall thin, outside grayish white pubescent, inside glabrous, with 6~9 rings. Nut ovoid or ellipsoid, glabrous, apex rounded. Fl. Jun., fr. Oct..

果   Fruit
摄影：杨庆松   Photo by: Yang Qingsong

枝叶   Branch and leaves
摄影：杨庆松   Photo by: Yang Qingsong

叶背   Leaf abaxial surface
摄影：杨庆松   Photo by: Yang Qingsong

径级分布表 DBH class

胸径区间 (Diameter class) (cm)	个体数 (No. of individuals in the plot)	比例 (Proportion) (%)
1~2	8	9.64
2~5	27	32.53
5~10	24	28.92
10~20	11	13.25
20~30	3	3.61
30~60	10	12.05
≥60	0	0.00

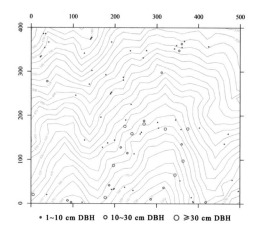

个体分布图 Distribution of individuals

18 云山青冈

Yún shān qīng gāng | Sessile Oak

Cyclobalanopsis sessilifolia (Blume) Schott.
壳斗科 | Fagaceae

代码 (SpCode) = CYCSES
个体数 (Individual number/20 hm^2) = 2459
最大胸径 (Max DBH) = 67.7 cm
重要值排序 (Importance value rank) = 5

常绿乔木，高达25 m。叶长椭圆形，全缘或近顶部具细齿，叶缘常反卷，两面近同色，无毛。花序轴被绒毛。壳斗杯状，被灰褐色短绒毛，具5～7环带，下部2～3环带有裂齿；坚果倒卵形或倒卵状椭圆形，果脐微凸。花期4～5月，果期10～11月。

Evergreen trees, up to 25 m tall. Leaf blade oblong-elliptic, margin entire or apically serrate; margin often revolute, subconcolorous, glabrous. Rachis of inflorescence tomentose. Cupule copular, outside grayish brown tomentose, with 5~7 rings, margin of basal 2~3 denticulate. Nut obovoid to obovoid-ellipsoid, slightly convex. Fl. Apr. - May, fr. Oct. - Nov..

树干　　　Trunk
摄影：杨庆松　Photo by: Yang Qingsong

花枝　　　Flowering Branch
摄影：杨庆松　Photo by: Yang Qingsong

果　　　Fruit
摄影：马遵平　Photo by: Ma Zunping

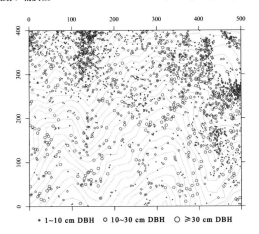

个体分布图　Distribution of individuals

径级分布表 DBH class

胸径等级 (Diameter class) (cm)	个体数 (No. of individuals in the plot)	比例 (Proportion) (%)
1～2	206	8.38
2～5	561	22.81
5～10	654	26.60
10～20	450	18.30
20～30	338	13.75
30～60	249	10.13
≥60	1	0.04

19 褐叶青冈

Hè yè qīng gāng | Brownleaf Oak

Cyclobalanopsis stewardiana (A. Camus) Y. C. Hsu et H. W. Jen
壳斗科 | Fagaceae

代码 (SpCode) = CYCSTE
个体数 (Individual number/20 hm^2) = 1
最大胸径 (Max DBH) = 2.9 cm
重要值排序 (Importance value rank) = 149

常绿乔木，高达12 m。小枝无毛。叶椭圆状披针形或长椭圆形，中部以上具锯齿，幼叶两面被绢毛，老叶近无毛，侧脉8～10对。花序轴及苞片被褐色绒毛。壳斗杯状，具5～9环带；坚果宽卵圆形，长0.8～1.5 cm，无毛，果脐凸起。花期7月，果期翌年10月。

Evergreen trees, up to 12 m tall. Branchlets glabrous. Leaf elliptic-lanceolate or oblong-elliptic, margin apical 1/2 shallowly serrate; sericeous hairy in both sides when young, glabrous when old, secondary veins 8~10 on each side of midvein. Rachis and bracts brown tomentose. Cupule copular, with 5~9 rings. Nut broadly ovoid, 0.8~1.5cm, glabrous, convex. Fl. Jul., fr. Oct. of following year.

树干　Trunk
摄影：杨庆松　Photo by: Yang Qingsong

枝叶　Branches and leaves
摄影：杨庆松　Photo by: Yang Qingsong

叶背　Leaf abaxial surface
摄影：杨庆松　Photo by: Yang Qingsong

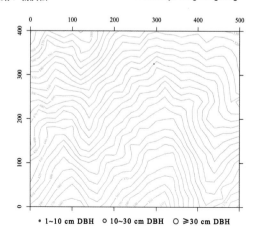

个体分布图　Distribution of individuals

径级分布表　DBH class

胸径区间 (Diameter class) (cm)	个体数 (No. of individuals in the plot)	比例 (Proportion) (%)
1～2	0	0.00
2～5	1	100.00
5～10	0	0.00
10～20	0	0.00
20～30	0	0.00
30～60	0	0.00
≥60	0	0.00

20 柯 (石栎)　　　　　　　　　　　　　　　Kē | Tanoak

Lithocarpus glaber (Thunb.) Nakai
壳斗科 | Fagaceae

代码 (SpCode) = LITGLA
个体数 (Individual number/20 hm^2) = 69
最大胸径 (Max DBH) = 25 cm
重要值排序 (Importance value rank) = 66

常绿乔木，高达15 m。小枝、幼叶柄、叶下面及花序轴均密被灰黄色短绒毛。叶革质或坚纸质，倒卵形、倒卵状椭圆形或长椭圆形，全缘或近顶端具浅齿；叶柄长1~2 cm。壳斗3 (5) 个成簇，碟状或浅碗状，无柄，高0.5~1 cm。坚果椭圆形，被白霜，果脐凹下。花期9~10月，果期翌年9~10月。

Evergreen trees, up to 15 m tall; branchlets, young petiole and rachis of inflorescences densely grayish yellow tomentose. Leaf leathery or papery, blade obvate, obvate-elliptic, or oblong, margin entire or shallow serrations on apical part, petiole 1~2 cm; cupules in clusters of 3(-5), plate to cupular, 5~10 mm tall. Nut ellipsoid, with white galucous, concave. Fl. Sep. - Oct., fr. Sep. - Oct. of following year.

花枝　　Flowering branch
摄影：杨庆松　Photo by: Yang Qingsong

枝叶　　Branches and leaves
摄影：杨庆松　Photo by: Yang Qingsong

果枝　　Fruiting branches
摄影：杨庆松　Photo by: Yang Qingsong

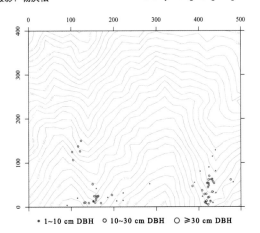

个体分布图　Distribution of individuals

径级分布表 DBH class

胸径等级 (Diameter class) (cm)	个体数 (No. of individuals in the plot)	比例 (Proportion) (%)
1~2	10	14.49
2~5	8	11.59
5~10	7	10.14
10~20	35	50.72
20~30	9	13.04
30~60	0	0.00
≥60	0	0.00

21 港柯 Gǎng kē | Harland Tanoak

Lithocarpus harlandii (Hance) Rehd.
壳斗科 | Fagaceae

代码 (SpCode) = LITHAR
个体数 (Individual number/20 hm^2) = 2656
最大胸径 (Max DBH) = 67.8 cm
重要值排序 (Importance value rank) = 7

常绿乔木，高达20 m。小枝被灰白色蜡鳞层，枝叶无毛。叶窄长椭圆形或披针状长椭圆形，全缘。果序轴被灰黄色毛。壳斗3个或成簇，密集，浅碗状，鳞片宽卵状三角形；坚果圆锥状，稍被白粉，果脐凹下，径1～1.5 cm。花期8～10月，果期翌年8～10月。

Evergreen trees, up to 18 m tall; branchlets grayish white wax scale layer, glabrous. Leaf blade narrow oblong or oblanceolate, margin entire; rachis of infructescences grayish yellow tomentose. Cupules in clusters or 3, dense, copular, bracts imbricate, broad ellipsoid, triangular. Nut conical, with white powdery, concave, 1~1.5 cm in diam. Fl. Aug. - Oct., fr. Aug. - Oct. of following year.

花枝　Flowering branch
摄影：杨庆松　Photo by: Yang Qingsong

幼苗　Seedling
摄影：杨庆松　Photo by: Yang Qingsong

果枝　Fruiting branches
摄影：杨海波　Photo by: Yang Haibo

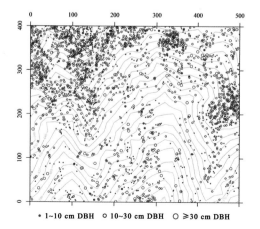
个体分布图　Distribution of individuals

径级分布表　DBH class

胸径区间 (Diameter class) (cm)	个体数 (No. of individuals in the plot)	比例 (Proportion) (%)
1～2	282	10.62
2～5	398	14.98
5～10	527	19.84
10～20	976	36.75
20～30	377	14.19
30～60	95	3.58
≥60	1	0.04

22 糙叶树

Cāo yè shù | Roughleaftree

Aphananthe aspera (Thunb.) Planch.
榆科 | Ulmaceae

代码 (SpCode) = APHASP
个体数 (Individual number/20 hm²) = 92
最大胸径 (Max DBH) = 66.8 cm
重要值排序 (Importance value rank) = 47

落叶乔木，高达25 m；树皮纵裂，粗糙。叶纸质，基部宽楔形或浅心形，两面粗糙，均有糙伏毛，基脉3出，侧脉直达叶缘。雄聚伞花序生于新枝的下部叶腋，雌花单生于新枝的上部叶腋。核果近球形，成熟时紫黑色。花期3~5月，果期8~10月。

Deciduous trees, up to 25 m tall. Bark longitudinally fissured, scabrous. Leaf papery, base broadly cuneate or shallow cordate; Leaf blade abaxially sparsely pubescent, adaxially scabrous with bristles. 3-veined from base, lateral pair extending to middle margin; Male flowers: in proximal leaf axil of young branchlets, Female flowers: solitary in distal leaf axil of young branchlets. Petiole flat tomentose. Drupes globose, ellipsoid or ovoid-globose. Fl. Mar. - May, fr. Aug. - Oct..

树干　　Trunk
摄影：杨庆松　Photo by: Yang Qingsong

雄花　　Male flowers
摄影：汪远　Photo by: Wang Yuan

果枝　　Fruiting branch
摄影：杨庆松　Photo by: Yang Qingsong

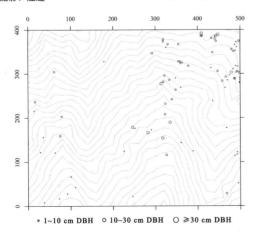

个体分布图　Distribution of individuals

径级分布表 DBH class

胸径等级 (Diameter class) (cm)	个体数 (No. of individuals in the plot)	比例 (Proportion) (%)
1~2	8	8.70
2~5	13	14.13
5~10	23	25.00
10~20	28	30.43
20~30	11	11.96
30~60	7	7.61
≥60	2	2.17

23 紫弹树

Zǐ dàn shù | Biond Nettletree

Celtis biondii Pamp.
榆科 | Ulmaceae

代码 (SpCode) = CELBIO
个体数 (Individual number/20 hm^2) = 55
最大胸径 (Max DBH) = 47.1 cm
重要值排序 (Importance value rank) = 73

落叶乔木，高达18 m。幼枝密被柔毛。冬芽黑褐色，芽鳞被柔毛，内层芽鳞的毛长而密。叶互生，卵形，基部稍偏斜，薄革质，中上部疏生浅齿，边稍反卷，上面脉纹多凹下。果序单生叶腋，总梗极短，果柄较长，梗连同果柄长1～2 cm，被糙毛。果近球形，径约5 mm，黄色或桔红色。花期4～5月，果期9～10月。

Deciduous trees, up to 18 m tall. Branchlets densely pubescent when young. Winter buds blackish brown, scales pubescent, hairs of intimal scales long and thick. Leaves alternate, ovate, base slightly oblique, thinly coriaceous, margin shallowly serrate on apical half, margin slightly revolute, adaxially veins usually concave. Infructescence solitary in distal leaf axil, common peduncle very short, carpopodium relatively long, peduncle and carpopodium long 1～2 cm, rough hairs. Drupe globose, 5 mm in diam approximately, yellow or reddish orange. Fl. Apr. - May, fr. Sep. - Oct..

树干　Trunk
摄影：杨庆松　Photo by: Yang Qingsong

枝叶　Branch and leaves
摄影：杨庆松　Photo by: Yang Qingsong

果枝　Fruiting branch
摄影：杨庆松　Photo by: Yang Qingsong

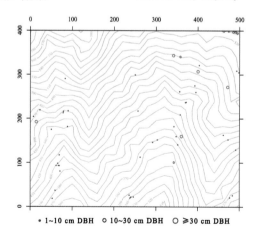
个体分布图 Distribution of individuals

径级分布表 DBH class

胸径区间 (Diameter class) (cm)	个体数 (No. of individuals in the plot)	比例 (Proportion) (%)
1～2	16	29.09
2～5	19	34.55
5～10	9	16.36
10～20	4	7.27
20～30	1	1.82
30～60	6	10.91
≥60	0	0.00

24 朴树　　Pǔ shù | China Nettletree

Celtis sinensis Pers.
榆科 | Ulmaceae

代码 (SpCode) = CELSIN
个体数 (Individual number/20 hm^2) = 25
最大胸径 (Max DBH) = 60 cm
重要值排序 (Importance value rank) = 65

落叶乔木，高达20 m。树皮灰白色，一年生枝密被柔毛。芽鳞无毛。叶卵形或卵状椭圆形，基部近对称或稍偏斜。核果单生叶腋，近球形，径5～7 mm，成熟时黄或橙黄色；果柄与叶柄近等长或稍短，被柔毛。花期3～4月，果期9～10月。

Deciduous trees, up to 20 m tall. Bark gray. Annual branch densely pubescent. Bud scales glabrous. Leaf ovate or ovate-elliptic, base symmetric or slightly oblique. Drupe distal leaf axil, globose, 5~7 mm in diam, yellow or orangish yellow when ripe; fruiting pedicel as long as petiole or relative short, pubescent. Fl. Mar. - Apr., fr. Sep. - Oct..

树干　　Trunk
摄影：杨庆松　Photo by: Yang Qingsong

枝叶　　Branch and leaves
摄影：杨庆松　Photo by: Yang Qingsong

果枝　　Fruiting branch
摄影：杨庆松　Photo by: Yang Qingsong

径级分布表 DBH class

胸径等级 (Diameter class) (cm)	个体数 (No. of individuals in the plot)	比例 (Proportion) (%)
1～2	2	8.00
2～5	3	12.00
5～10	6	24.00
10～20	4	16.00
20～30	4	16.00
30～60	5	20.00
≥60	1	4.00

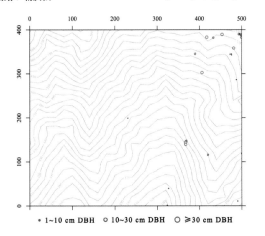

个体分布图　Distribution of individuals

25 西川朴 | Xī chuān pǔ | W. Sichuan Nettletree

Celtis vandervoetiana Schneid.
榆科 | Ulmaceae

代码 (SpCode) = CELVAN
个体数 (Individual number/20 hm^2) = 92
最大胸径 (Max DBH) = 54.6 cm
重要值排序 (Importance value rank) = 57

落叶乔木，高达20 m。树皮灰色，一年生枝红褐色，无毛。叶片近革质，基部稍偏斜，边缘近基部或中部以上有粗锯齿，脉均隆起。叶柄长1.2~2.0 cm，无毛。核果单生叶腋，卵状椭圆形，果梗长1.7~3.3 cm。花期3~4月，果期9~10月。

Deciduous trees, up to 20 m tall. Bark gray, annual branch reddish brown, glabrous. Leaf blade nearly leathery, base slightly oblique, margin roughly serrate on base or apical 1/2, veins raised. Petiole 1.2~2 cm, glabrous. Drupe distal leaf axil, ovate-elliptic, carpopodium 1.7~3.3 cm. Fl. Mar.-Apr., fr. Sep. - Oct..

树干　　　　Trunk
摄影：杨庆松　Photo by: Yang Qingsong

枝叶　　　　Branch and leaves
摄影：杨庆松　Photo by: Yang Qingsong

果枝　　　　Fruiting branch
摄影：杨庆松　Photo by: Yang Qingsong

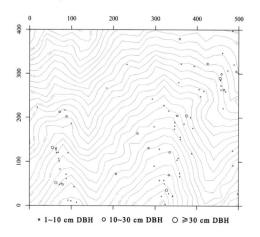

个体分布图 Distribution of individuals

径级分布表 DBH class

胸径区间 (Diameter class) (cm)	个体数 (No. of individuals in the plot)	比例 (Proportion) (%)
1~2	22	23.91
2~5	31	33.70
5~10	20	21.74
10~20	10	10.87
20~30	4	4.35
30~60	5	5.43
≥60	0	0.00

26 山油麻

Shān yóu má | Diels Wildjute

Trema cannabina var. *dielsiana* (Hand.-Mazz.) C. J. Chen
榆科 | Ulmaceae

代码 (SpCode) = TRECAN
个体数 (Individual number/20 hm^2) = 37
最大胸径 (Max DBH) = 5.4 cm
重要值排序 (Importance value rank) = 98

落叶灌木或小乔木；小枝纤细，黄绿色，密被斜伸粗毛。叶薄纸质，叶面绿色，上面被糙毛，下面密被柔毛，脉上被粗毛，基部有明显的三出脉。花单性，雌雄同株，雌花序常生于花枝的上部叶腋，雄花序常生于花枝的下部叶腋。核果近球形或阔卵圆形，熟时桔红色。花期3～6月，果期9～10月。

Deciduous shrubs or small trees; branchlets fine, yellowish green, densely hirsute with obliquely spreading hairs. Leaf blade thin papery, adaxially green, rough hairs, abaxially densely pubescent, vein shag, 3-veined from base obviously. Flowers unisexual, monoecism. Female inflorescence usually grow in distal leaf axil of spray; male inflorescence usually grow in proximal leaf axil of spray. Drupes globose or broadly ovoid, reddish orange when ripe. Fl. Mar. - Jun., fr. Sep. - Oct..

叶　　Leaves
摄影：杨庆松　Photo by: Yang Qingsong

果枝　　Fruiting branch
摄影：杨庆松　Photo by: Yang Qingsong

枝叶　　Branch and leaves
摄影：杨庆松　Photo by: Yang Qingsong

径级分布表 DBH class

胸径等级 (Diameter class) (cm)	个体数 (No. of individuals in the plot)	比例 (Proportion) (%)
1～2	15	40.54
2～5	20	54.05
5～10	2	5.41
10～20	0	0.00
20～30	0	0.00
30～60	0	0.00
≥60	0	0.00

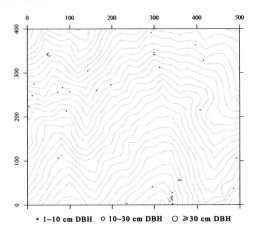

个体分布图 Distribution of individuals

27 杭州榆   Háng zhōu yú | Chang Elm

Ulmus changii W. C. Cheng
榆科 | Ulmaceae

代码 (SpCode) = ULMCHA
个体数 (Individual number/20 hm^2) = 129
最大胸径 (Max DBH) = 38 cm
重要值排序 (Importance value rank) = 60

落叶乔木，高达20 m。树皮灰褐色，幼枝密被毛。叶卵形或卵状椭圆形，侧脉12~20 (24) 对，常具单锯齿，稀兼具或全为重锯齿；叶柄长3~8 mm。翅果长圆形或椭圆状长圆形，稀近圆形，长1.5~3.5 cm，被短毛。果核位于翅果中部或稍下；果柄密被短毛。花果期3~4月。

Deciduous trees, up to 20 m tall. Bark grayish brown. Branchlets densely pubescent when young. Leaf ovate or ovate-elliptic, secondary veins 12~20(-24) on each side of midvein, margin simply serrate or rarely doubly serrate; petiole 3-8 mm. Samaras oblong-elliptic, rare round, 1.5~3.5 cm, carpopodium densely pubescent. Fl. and fr. Mar. - Apr..

树干　　　　　Trunk
摄影：杨庆松　　Photo by: Yang Qingsong

枝叶　　　　　Branch and leaves
摄影：杨庆松　　Photo by: Yang Qingsong

老叶　　　　　Old leaf
摄影：杨庆松　　Photo by: Yang Qingsong

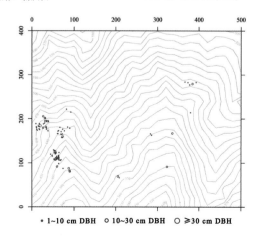

个体分布图 Distribution of individuals

径级分布表 DBH class

胸径区间 (Diameter class) (cm)	个体数 (No. of individuals in the plot)	比例 (Proportion) (%)
1~2	25	19.38
2~5	59	45.74
5~10	28	21.71
10~20	8	6.20
20~30	7	5.43
30~60	2	1.55
≥60	0	0.00

28 大叶榉树

Dà yè jǔ shù | Bigleaf Waterelm

Zelkova schneideriana Hand.-Mazz.
榆科 | Ulmaceae

代码 (SpCode) = ZELSCH
个体数 (Individual number/20 hm^2) = 17
最大胸径 (Max DBH) = 53.7 cm
重要值排序 (Importance value rank) = 84

落叶乔木，高达35 m。树皮灰褐色至深灰色，呈不规则片状剥落。一年生枝密被伸展灰色柔毛；冬芽常2个并生。叶卵形或椭圆状披针形，基部稍偏斜，具圆齿状锯齿，侧脉8～15对。雄花1～3朵生于叶腋，雌花或两性花常单生于幼枝上部叶腋。核果上部歪斜。花期4月，果期9～11月。

Deciduous trees, up to 35 m; bark grayish brown to dark gray, irregular exfoliating. Annual branch densely covered with gray pubescence; winter buds usually two united. Leaf ovate to elliptic-lanceolate, base slightly oblique, margin serrate to crenate, secondary veins 8~15 on each side of midvein. Male flowers: 1~3 clustered in leaf axil. Female and bisexual flowers usually solitary in distal leaf axil of young branchlets. Drupes. Fl. Apr., fr. Sep. - Nov..

树皮　　　　　　Bark
摄影：杨庆松　　Photo by: Yang Qingsong

枝叶　　　　　　Branch and leaves
摄影：杨庆松　　Photo by: Yang Qingsong

叶　　　　　　　Leaf
摄影：杨庆松　　Photo by: Yang Qingsong

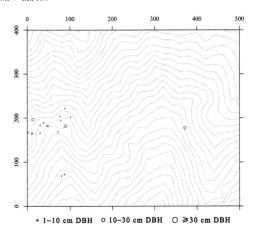

个体分布图　Distribution of individuals

径级分布表 DBH class

胸径等级 (Diameter class) (cm)	个体数 (No. of individuals in the plot)	比例 (Proportion) (%)
1～2	5	29.41
2～5	6	35.29
5～10	0	0.00
10～20	1	5.88
20～30	2	11.76
30～60	3	17.65
≥60	0	0.00

29 矮小天仙果

Ai xiǎo tiān xiān guǒ | Beechey Fig

Ficus erecta Thunb.
桑科 | Moraceae

代码 (SpCode) = FICERE
个体数 (Individual number/20 hm^2) = 3
最大胸径 (Max DBH) = 7.2 cm
重要值排序 (Importance value rank) = 128

落叶小乔木或灌木状，高达3~4 m；树皮灰褐色。小枝密被硬毛。叶厚纸质，倒卵状椭圆形，长7~20 cm，先端短渐尖，基部圆或浅心形；叶柄长1~4 cm，纤细，密被灰白色硬毛。榕果单生叶腋，具总柄，球形或近梨形，成熟时红色。雌雄异株。花果期5~6月。

Deciduous small trees or shrubs, 3~4 m tall. Bark grayish brown. Branchlets densely hirsute. Leaf thickly papery, blade obovate-elliptic, 7~20 cm long, apex shortly acuminate, base rounded or shallow cordate. Petiole slender, 1~4 cm, densely grayish white hirsute. Figs axillary on normal leafy shoots, with phyllomophore, globose or pyriform. Dioecism. Fl. May - Jun., fr. May - Jun..

树干　　Trunk
摄影：杨庆松　Photo by: Yang Qingsong

叶　　Leaves
摄影：杨庆松　Photo by: Yang Qingsong

果枝　　Fruiting branches
摄影：杨庆松　Photo by: Yang Qingsong

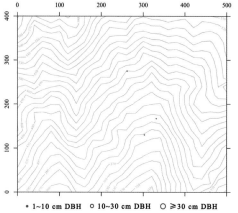

个体分布图 Distribution of individuals

径级分布表 DBH class

胸径区间 (Diameter class) (cm)	个体数 (No. of individuals in the plot)	比例 (Proportion) (%)
1~2	0	0.00
2~5	1	33.33
5~10	2	66.67
10~20	0	0.00
20~30	0	0.00
30~60	0	0.00
≥60	0	0.00

30 构棘（葨芝） Gòu jí | Vietnam Cudrania

Maclura cochinchinensis (Lour.) Corner
桑科 | Moraceae

代码 (SpCode) = MACCOC
个体数 (Individual number/20 hm^2) = 2
最大胸径 (Max DBH) = 1.4 cm
重要值排序 (Importance value rank) = 143

常绿直立或蔓生灌木，高2～4 m。树皮灰褐色，略粗糙；分枝无毛，皮孔散生，具直立或略弯的枝刺。叶片革质，椭圆状披针形到长圆形，不裂，先端钝或渐尖或有凹缺，基部楔形，全缘，两面无毛；叶柄长5～10 mm。聚花果球形，肉质，橙红色。花期4～5月，果期6～7月。

Evergreen shrubs, erect or scandent, 2~4 m tall. Bark gray, tough, Branches glabrous, scattered lenticels; spines curved or straight. Leaf leathery, blade elliptic-lanceolate to oblong, glabrous, base cuneate, margin entire, apex rounded to shortly acuminate; petiole ca. 0.5~1 cm. Fruiting syncarp reddish orange when mature, ovoid, succulent. Fl. Apr. - May, fr. Jun. - Jul..

茎　　Stem
摄影：杨庆松　Photo by: Yang Qingsong

枝叶　　Branch and leaves
摄影：杨庆松　Photo by: Yang Qingsong

果枝　　Fruiting branch
摄影：葛斌杰　Photo by: Ge Binjie

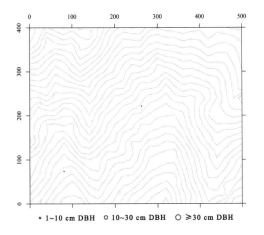

个体分布图　Distribution of individuals

径级分布表 DBH class

胸径等级 (Diameter class) (cm)	个体数 (No. of individuals in the plot)	比例 (Proportion) (%)
1～2	2	100.00
2～5	0	0.00
5～10	0	0.00
10～20	0	0.00
20～30	0	0.00
30～60	0	0.00
≥60	0	0.00

31 柘 Zhè | Tricuspid Cudrania

Maclura tricuspidata Carrière
桑科 | Moraceae

代码 (SpCode) = MACTRI
个体数 (Individual number/20 hm^2) = 1
最大胸径 (Max DBH) = 5.3 cm
重要值排序 (Importance value rank) = 146

落叶小乔木，常为灌木状，1~7 m 高。树皮淡灰色，成不规则的薄片剥落。小枝无毛，刺长0.5~2 cm。叶卵形或菱状卵形，先端渐尖，基部楔形或圆，全缘或3裂，两面无毛，或下面被柔毛，侧脉4~6对。聚花果近球形，径约2.5 cm，肉质，熟时桔红色。花期5~6月，果期6~7月。

Deciduous small trees, usually shrub, 1~7 m tall. Bark light grey, irregular exfoliating. Branchlets glabrous, spines 0.5~2 cm. Leaf blade ovate to rhombic-ovate, apex acuminate, base cuneate or rounded, margin entire or 3-lobed, glabrous in both sides or abaxially pubescent, secondary veins 4~6 on each side of midvein. Fruiting syncarp globose, 2.5 cm in diam approximately, succulent, orange red when mature. Fl. May - Jun., fr. Jun. - Jul.

树干　Trunk
摄影：杨庆松　Photo by: Yang Qingsong

枝叶　Branch and leaves
摄影：杨庆松　Photo by: Yang Qingsong

果枝　Fruiting branches
摄影：汪远　Photo by: Wang Yuan

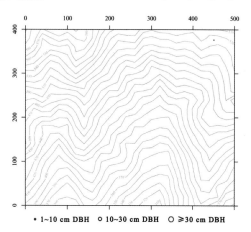
个体分布图　Distribution of individuals

径级分布表　DBH class

胸径区间 (Diameter class) (cm)	个体数 (No. of individuals in the plot)	比例 (Proportion) (%)
1~2	0	0.00
2~5	0	0.00
5~10	1	100.00
10~20	0	0.00
20~30	0	0.00
30~60	0	0.00
≥60	0	0.00

32 紫麻

Zǐ má | Shrubby Woodnettle

Oreocnide frutescens (Thunb.) Miq.
荨麻科 | Urticaceae

代码 (SpCode) = OREFRU
个体数 (Individual number/20 hm^2) = 3
最大胸径 (Max DBH) = 1.4 cm
重要值排序 (Importance value rank) = 136

落叶灌木，1～3 m 高。叶卵形或窄卵形，稀倒卵形，草质，常生于枝上部，长3～15 cm，先端渐尖或尾尖，下面常被灰白色毡毛，后渐脱落，基出3脉，侧脉2～3对；叶柄长1～7 cm。花序生于去年生枝和老枝，几无梗，呈簇生状。瘦果卵球状。花期3～5月，果期6～10月。

Deciduous shrubs, 1~3 m tall. Leaf blade ovate or narrowly ovate, rarely obovate, herbaceous, usually grow in top of branch, 3~15 cm long, apex acuminate or caudate-acuminate, base often grayish white tomentose, then gradually fall, basinerved 3, lateral veins 2~3 pairs. Petiole 1~7 cm. Inflorescences produced with or before new leaf flush in axils of fallen leaves and older branches, almost sessile clusters. Achene ovoid. Fl. Mar. - May, fr. Jun. - Oct..

花枝　　Flowering branch
摄影：杨庆松　Photo by: Yang Qingsong

叶　　Leaves
摄影：杨庆松　Photo by: Yang Qingsong

种子　　Seeds
摄影：杨庆松　Photo by: Yang Qingsong

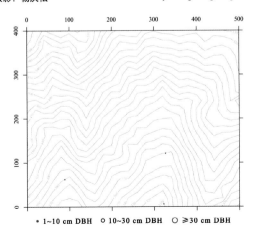
个体分布图 Distribution of individuals

径级分布表 DBH class

胸径等级 (Diameter class) (cm)	个体数 (No. of individuals in the plot)	比例 (Proportion) (%)
1～2	3	100.00
2～5	0	0.00
5～10	0	0.00
10～20	0	0.00
20～30	0	0.00
30～60	0	0.00
≥60	0	0.00

33 青皮木

Qīng pí mù | Common Greentwig

Schoepfia jasminodora Sieb. et Zucc.
铁青树科 | Olacaceae

代码 (SpCode) = SCHJAS
个体数 (Individual number/20 hm^2) = 101
最大胸径 (Max DBH) = 55.1 cm
重要值排序 (Importance value rank) = 55

落叶小乔木，高3～15 m；树皮灰褐色；具短枝，新枝自去年生短枝上抽出，嫩时红色，老枝灰褐色，小枝干后栗褐色。叶纸质，卵形或长卵形；叶柄长2～3 mm，红色。花无梗，2(3)～9朵排成穗状花序状的螺旋状聚伞花序；花冠钟形或宽钟形，白色或浅黄色。花叶同放。花期3～5月，果期4～6月。

Deciduous small trees, 3~15 m tall. Bark grayish brown. With short shoots, new branchlets produced in succession from last year short shoots, red when young, grayish brown when old, chestnut when dry. Leaf blade papery, ovate or oblong-ovate. Petiole 2~3 mm, red. Inflorescences 2(~3) ~9 arranged spadicose ataxinomic cyme. Corolla campaniform or broadly campaniform, white or pale yellow. Flowers opening at same time as leaves. Fl. Mar. - May, fr. Apr. - Jun..

枝叶　Branch and leaves
摄影：王樟华　Photo by: Wang Zhanghua

花枝　Flowering branches
摄影：杨庆松　Photo by: Yang Qingsong

果枝　Fruiting branches
摄影：杨庆松　Photo by: Yang Qingsong

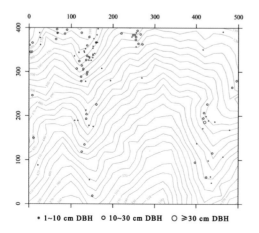

个体分布图　Distribution of individuals

径级分布表　DBH class

胸径区间 (Diameter class) (cm)	个体数 (No. of individuals in the plot)	比例 (Proportion) (%)
1～2	11	10.89
2～5	12	11.88
5～10	29	28.71
10～20	35	34.65
20～30	13	12.87
30～60	1	0.99
≥60	0	0.00

34 南天竹　　　　　　　　　　　　　　　　　　Nán tiān zhú | Common Nandina

Nandina domestica Thunb.
小檗科 | Berberidaceae

代码 (SpCode) = NANDOM
个体数 (Individual number/20 hm^2) = 2
最大胸径 (Max DBH) = 1.6 cm
重要值排序 (Importance value rank) = 142

常绿小灌木。茎常丛生而少分枝，高1～3 m，光滑无毛，幼枝常为红色，老后呈灰色。叶互生，集生于茎的上部，三回羽状复叶，长30～50 cm；二至三回羽片对生；小叶薄革质，椭圆形或椭圆状披针形，全缘，上面深绿色，冬季变红色。圆锥花序。浆果球形。种子扁圆形。花期3～6月，果期5～11月。

Evergreen small shrubs. Stems often clump and subramous, up to 1~3 m tall, glabrous, branchlets reddish when young. Leaves alternate, set in the upper portion of the stem, tripinnate leaf, 30~50 cm. Pinnule or tripinnate opposite. Leaflets weakly leathery, elliptic or elliptic-lanceolate, margin entire, adaxially dark green, reddish when winter. Panicle. Berry sphere. Seeds oblate-spheroidal. Fl. Mar. - Jun., fr. May - Nov..

枝叶　Branch and leaves
摄影：杨庆松　Photo by: Yang Qingsong

花枝　Flowering branches
摄影：杨庆松　Photo by: Yang Qingsong

果枝　Fruiting branches
摄影：杨庆松　Photo by: Yang Qingsong

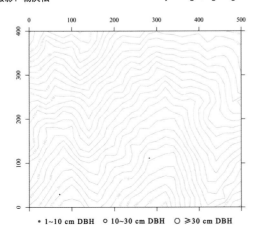

个体分布图　Distribution of individuals

径级分布表　DBH class

胸径等级 (Diameter class) (cm)	个体数 (No. of individuals in the plot)	比例 (Proportion) (%)
1～2	2	100.00
2～5	0	0.00
5～10	0	0.00
10～20	0	0.00
20～30	0	0.00
30～60	0	0.00
≥60	0	0.00

35 天目木兰

Tiān mù mù lán | Tianmu Magnolia

Magnolia amoena W. C. Cheng
木兰科 | Magnoliaceae

代码 (SpCode) = MAGAMO
个体数 (Individual number/20 hm^2) = 4
最大胸径 (Max DBH) = 28.3 cm
重要值排序 (Importance value rank) = 111

落叶乔木，高达12 m，树皮灰色或灰白色；芽被灰白色紧贴毛；嫩枝绿色，老枝带紫色。叶纸质，宽倒披针形，倒披针状椭圆形，长10～15 cm，宽3.5～5 cm。聚合果圆柱形，长4～10 cm。花期4～5月，果期9～10月。

Deciduous trees, up to 12 m tall. Bark grey or grayish white. Buds grayish white clinging pubescent. Branchlets green when young but purple when mature. Leaves papery, broadly oblanceolate or oblanceolate-elliptic, 10~15 × 3.5~5 cm. Aggregate fruits cylindrical, 4~10 cm. Fl. Apr. - May, fr. Sep. - Oct..

树干   Trunk
摄影：杨庆松   Photo by: Yang Qingsong

枝叶   Branch and leaves
摄影：杨庆松   Photo by: Yang Qingsong

叶   Leaf
摄影：杨庆松   Photo by: Yang Qingsong

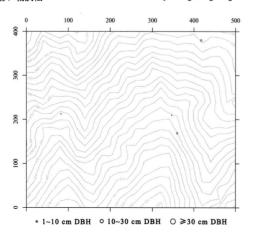

个体分布图 Distribution of individuals

径级分布表 DBH class

胸径区间 (Diameter class) (cm)	个体数 (No. of individuals in the plot)	比例 (Proportion) (%)
1～2	0	0.00
2～5	1	25.00
5～10	1	25.00
10～20	1	25.00
20～30	1	25.00
30～60	0	0.00
≥60	0	0.00

36 红毒茴 (披针叶茴香)　　　　　　　　　Hóng dú huí | Poisonous Eightangle

Illicium lanceolatum A. C. Smith
八角科 | Illiciaceae

代码 (SpCode) = ILLLAN
个体数 (Individual number/20 hm²) = 1434
最大胸径 (Max DBH) = 37.6 cm
重要值排序 (Importance value rank) = 25

常绿小乔木，高3～10 m；枝条纤细，树皮浅灰色至灰褐色。叶互生或稀疏地簇生于小枝顶端或排成假轮生，革质，披针形、倒披针形或倒卵状椭圆形，网脉不明显；叶柄纤细，长7～15 mm。花腋生或近顶生，红色，深红色。果梗长可达6 cm，纤细，蓇葖10～14枚(少有9)轮状排列。花期4～6，果期8～10月。

Evergreen small trees, up to 3~10 m tall. Branchlets slender, bark light grey to grayish brown. Leaves alternate, sparsely fascioled on apex of branchlets or polytomous, leathery, lanceolate, oblanceolate or obovate-elliptic, vein not obvious. Petiole slender, 7~15 mm. Flowers axillary or subterminal, red to dark red. Fruit peduncle 6 cm, slender, with 10~14 (rare 9) follicles rotate arrange. Fl. Apr. - Jun., fr. Aug. - Oct..

幼苗　Seedling
摄影：刘何铭　Photo by: Liu Heming

花枝　Flowering branches
摄影：杨庆松　Photo by: Yang Qingsong

花　Flower
摄影：杨庆松　Photo by: Yang Qingsong

径级分布表 DBH class

胸径等级 (Diameter class) (cm)	个体数 (No. of individuals in the plot)	比例 (Proportion) (%)
1～2	459	32.01
2～5	594	41.42
5～10	254	17.71
10～20	101	7.04
20～30	25	1.74
30～60	1	0.07
≥60	0	0.00

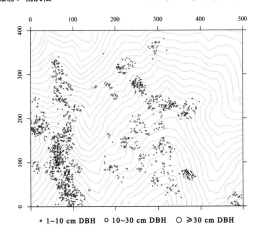

个体分布图　Distribution of individuals

37 樟   Zhāng | Camphortree

Cinnamomum camphora (Linn.) Presl
樟科 | Lauraceae

代码 (SpCode) = CINCAM
个体数 (Individual number/20 hm^2) = 1
最大胸径 (Max DBH) = 1.0 cm
重要值排序 (Importance value rank) = 154

常绿乔木，高达30 m；枝、叶及木材均有樟脑气味；树皮有不规则的纵裂。叶互生，卵状椭圆形，长6~12 cm，宽2.5~5.5 cm，具离基三出脉，侧脉及支脉脉腋上面明显隆起下面有明显腺窝。圆锥花序腋生，具梗。果卵球形或近球形，直径6~8 mm，熟时紫黑色。花期4~5月，果期8~11月。

Evergreen trees, up to 30 m tall. Whole plant strongly camphorscented; bark irregularly and longitudinally fissured. Leaves alternate, ovate-elliptic, 6~12 × 2.5~5.5 cm, triplinerved, axils of lateral veins and veins conspicuously dome-shaped. Panicle axillary, peduncle. Fruit ovoid or subglobose, 6~8 mm in diam, purple-black when mature. Fl. Apr. - May, fr. Aug. - Nov..

叶   Leaves
摄影：杨庆松   Photo by: Yang Qingsong

花枝   Flowering branches
摄影：杨庆松   Photo by: Yang Qingsong

果枝   Fruiting branches
摄影：杨庆松   Photo by: Yang Qingsong

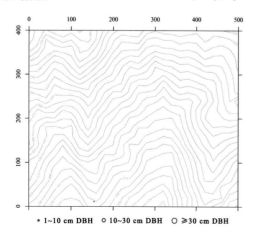
个体分布图   Distribution of individuals

径级分布表   DBH class

胸径区间 (Diameter class) (cm)	个体数 (No. of individuals in the plot)	比例 (Proportion) (%)
1~2	1	100.00
2~5	0	0.00
5~10	0	0.00
10~20	0	0.00
20~30	0	0.00
30~60	0	0.00
≥60	0	0.00

38 香桂

Xiāng guì | Fragrant Cinnamon

Cinnamomum subavenium Miq.
樟科 | Lauraceae

代码 (SpCode) = CINSUB
个体数 (Individual number/20 hm^2) = 1214
最大胸径 (Max DBH) = 44.6 cm
重要值排序 (Importance value rank) = 21

常绿乔木，高达20 m；树皮灰色，平滑。叶互生或近对生，革质，叶片椭圆形、卵状椭圆形至卵状披针形，先端急尖至渐尖，基部圆形或楔形。三出脉或近离基三出脉，侧脉斜上升，直贯叶端，网脉两面不明显。圆锥花序腋生。果椭圆形，长约7 mm，熟时蓝黑色。花期6～7月，果期8～10月。

Evergreen trees, up to 20 m tall. Bark gray, smooth. Leaves alternate or subopposite, leathery, elliptic or ovate-elliptic to ovate-lanceolate, apex acuminate or acute, base rounded to cuneate, trinerved or triplinerved, secondary veins oblique and extending to leaf apex, lateral veins inconspicuously. Panicle axillary. Fruit ellipsoid, 7 mm, blue-black when mature. Fl. Jun. - Jul., fr. Aug. - Oct..

树干　　Trunk
摄影：杨庆松　Photo by: Yang Qingsong

幼枝　　Young branch
摄影：杨庆松　Photo by: Yang Qingsong

叶　　Leaf
摄影：杨庆松　Photo by: Yang Qingsong

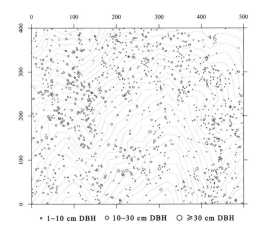

个体分布图　Distribution of individuals

径级分布表 DBH class

胸径等级 (Diameter class) (cm)	个体数 (No. of individuals in the plot)	比例 (Proportion) (%)
1～2	459	37.81
2～5	291	23.97
5～10	186	15.32
10～20	222	18.29
20～30	44	3.62
30～60	12	0.99
≥60	0	0.00

39 红果山胡椒 (红果钓樟)   Hóng guǒ shān hú jiāo | Redfruit Spicebush

Lindera erythrocarpa Makino
樟科 | Lauraceae

代码 (SpCode) = LINERY
个体数 (Individual number/20 hm^2) = 4
最大胸径 (Max DBH) = 11.5 cm
重要值排序 (Importance value rank) = 121

落叶灌木至小乔木，高可达6.5 m。树皮灰褐色，小枝皮孔多数，显著隆起。叶互生，纸质，叶片倒披针形至倒卵状披针形，先端渐尖，基部狭楔形下延，上面绿色，下面灰白色。伞形花序位于腋芽两侧。果球形，熟时红色。花期4月，果期9～10月。

Deciduous shrubs or small trees, up to 6.5 m tall. Bark graybrown. Young branchlets usually gray-white or gray-yellow, many lenticellate, scabrous from corky protuberances. Leaves alternate, papery, leaves blade oblanceolate or obovate-lanceolate, apex acuminate, base narrow cuneate, green adaxially, gray-white abaxially. Umbels in both side of axillary. Fruits globose, red when mature. Fl. Apr., fr. Sep. - Oct..

树干 Trunk
摄影：杨庆松 Photo by: Yang Qingsong

叶 Leaves
摄影：杨庆松 Photo by: Yang Qingsong

果 Fruits
摄影：杨庆松 Photo by: Yang Qingsong

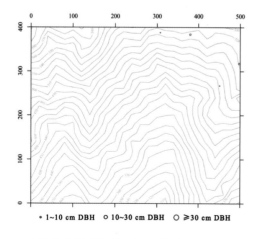
个体分布图 Distribution of individuals

径级分布表 DBH class

胸径区间 (Diameter class) (cm)	个体数 (No. of individuals in the plot)	比例 (Proportion) (%)
1～2	0	0.00
2～5	1	25.00
5～10	1	25.00
10～20	2	50.00
20～30	0	0.00
30～60	0	0.00
≥60	0	0.00

40 山胡椒

Shān hú jiāo | Greyblue Spicebush

Lindera glauca (Sieb. et Zucc.) Blume
樟科 | Lauraceae

代码 (SpCode) = LINGLA
个体数 (Individual number/20 hm²) = 53
最大胸径 (Max DBH) = 7.1 cm
重要值排序 (Importance value rank) = 94

落叶灌木或小乔木，高可达8 m；树皮平滑，灰色或灰白色。叶互生，宽椭圆形、椭圆形、倒卵形到狭倒卵形，上面深绿色，下面淡绿色，被白色柔毛，纸质，羽状脉；叶枯后不落，翌年新叶发出时落下。果球形，直径6~7 mm，熟时黑褐色。花期3~4月，果期7~8月。

Deciduous shrubs or small trees, up to 8 m tall. Bark smooth, gray or gray-white. Leaves alternate, broadly elliptic, elliptic, obovate to narrowly obovate, dark green adaxially, light green abaxially, white pubescent, papery, pinninerved. Leaves fall when young leaves grow out the next year. Fruits globose, 6~7 mm in diam, black brown when mature. Fl. Mar. - Apr., fr. Jul. - Aug..

枝叶　　Branch and leaves
摄影：杨庆松　　Photo by: Yang Qingsong

花枝　　Flowering branch
摄影：杨庆松　　Photo by: Yang Qingsong

果枝　　Fruiting branch
摄影：杨庆松　　Photo by: Yang Qingsong

径级分布表 DBH class

胸径等级 (Diameter class) (cm)	个体数 (No. of individuals in the plot)	比例 (Proportion) (%)
1~2	11	20.75
2~5	31	58.49
5~10	11	20.75
10~20	0	0.00
20~30	0	0.00
30~60	0	0.00
≥60	0	0.00

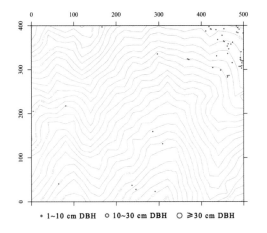

个体分布图 Distribution of individuals

41 红脉钓樟

Hóng mài diào zhāng | Redvein Spicebush

Lindera rubronervia Gamble
樟科 | Lauraceae

代码 (SpCode) = LINRUB
个体数 (Individual number/20 hm^2) = 342
最大胸径 (Max DBH) = 13.9 cm
重要值排序 (Importance value rank) = 45

落叶灌木或小乔木，高达5 m；树皮黑灰色，有皮孔。叶互生，卵形，离基三出脉，通常在中脉中部以上侧脉每边3～4条，脉和叶柄秋后变为红色；叶柄长5～10 mm，被短柔毛。伞形花序腋生。果近球形，直径1 cm。花期3～4月，果期8～9月。

Deciduous shrubs or small trees, up to 5 m tall. Bark dark gray, with lenticel. Leaves alternate, ovate, triplinerved, 3~4 lateral veins at both sides of midrib above the middle, vein and petiole become red when autumn. Petiole 5~10 mm, pubescent. Umbel axillary. Fruit subglobose, 1 cm in diam. Fl. Mar. - Apr., fr. Aug. - Sep..

幼苗　Seedling
摄影：刘何铭　Photo by: Liu Heming

果枝　Fruiting branch
摄影：汪远　Photo by: Wang Yuan

老叶　Old leaves
摄影：杨庆松　Photo by: Yang Qingsong

径级分布表 DBH class

胸径区间 (Diameter class) (cm)	个体数 (No. of individuals in the plot)	比例 (Proportion) (%)
1～2	78	22.81
2～5	163	47.66
5～10	84	24.56
10～20	17	4.97
20～30	0	0.00
30～60	0	0.00
≥60	0	0.00

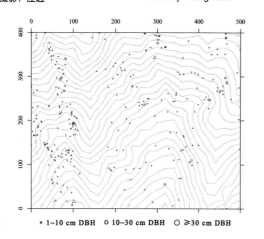

个体分布图 Distribution of individuals

42 豹皮樟

Bào pí zhāng | Leopard Skin Litsea

Litsea coreana var. *sinensis* (Allen) Yang et P. H. Huang
樟科 | Lauraceae

代码 (SpCode) = LITCOR
个体数 (Individual number/20 hm^2) = 306
最大胸径 (Max DBH) = 29.9 cm
重要值排序 (Importance value rank) = 44

常绿乔木，高8～15 m；树皮灰色，呈小鳞片状剥落，脱落后呈鹿皮斑痕。小枝无毛。叶互生，长圆形或披针形，先端急尖，基部楔形，革质，两面无毛，羽状脉，侧脉每边7～10条；叶柄长6～16 mm，上面有柔毛。伞形花序腋生。果近球形，直径7～8 mm。花期8～9月，果期翌年夏季。

Evergreen trees, 8~15 m tall. Bark gray, scaly peeled off, with collated spots exposed. Branchlets glabrous. Leaves alternate; leaf blade oblong or lanceolate, glabrous on both surfaces, lateral veins 7~10 pairs, base cuneate, apex shortly acuminate. Umbels axillary. Fruit subglobose, 7~8 mm in diam. Fl. Aug. - Sep., fr. summer of next year.

树干　　Trunk
摄影：杨庆松　Photo by: Yang Qingsong

叶　　Leaves
摄影：杨庆松　Photo by: Yang Qingsong

花枝　　Flowering branch
摄影：汪远　Photo by: Wang Yuan

径级分布表 DBH class

胸径等级 (Diameter class) (cm)	个体数 (No. of individuals in the plot)	比例 (Proportion) (%)
1～2	34	11.11
2～5	147	48.04
5～10	93	30.39
10～20	29	9.48
20～30	3	0.98
30～60	0	0.00
≥60	0	0.00

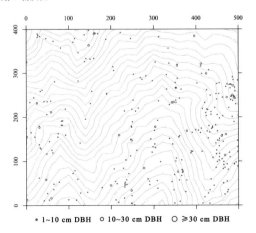

个体分布图　Distribution of individuals

43 山鸡椒（山苍子）   Shān jī jiāo | Mountain Spicy Tree

Litsea cubeba (Lour.) Pers.
樟科 | Lauraceae

代码 (SpCode) = LITCUB

个体数 (Individual number/20 hm^2) = 98

最大胸径 (Max DBH) = 17.2 cm

重要值排序 (Importance value rank) = 81

落叶小乔木，高达8～10 m，幼树树皮黄绿色，光滑，老树树皮灰褐色。小枝细长，绿色，无毛，枝、叶具芳香味。叶互生，披针形或长圆形，纸质。伞形花序单生或簇生；总梗细长，长6～10 mm；先叶开放。果近球形，直径约5 mm，幼时绿色，成熟时黑色。花期2～3月，果期7～8月。

Deciduous small trees, up to 8~10 m tall. Bark yellowish green and smooth when young but grayish brown when mature. Branchlets tenuous, green, glabrous, branchlets and leaves with aromatic flavor. Leaves alternate, lanceolate or oblong, papery. Umbel solitary or clustered, peduncle tenuous, 6~10 mm, appearing before leaves. Fruit subglobose, ca. 5 mm in diam, green when young, black at maturity. Fl. Feb. - Mar., fr. Jul. - Aug..

叶   Leaves
摄影：杨庆松   Photo by: Yang Qingsong

花枝   Flowering branches
摄影：杨庆松   Photo by: Yang Qingsong

果枝   Fruiting branch
摄影：杨庆松   Photo by: Yang Qingsong

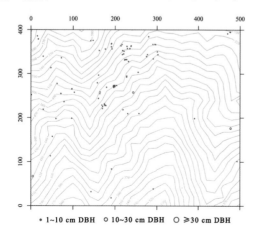
个体分布图   Distribution of individuals

径级分布表 DBH class

胸径区间 (Diameter class) (cm)	个体数 (No. of individuals in the plot)	比例 (Proportion) (%)
1～2	55	56.12
2～5	34	34.69
5～10	6	6.12
10～20	3	3.06
20～30	0	0.00
30～60	0	0.00
≥60	0	0.00

44 黄丹木姜子

Huáng dān mù jiāng zǐ | Elongate Litse

Litsea elongata (Nees) J. D. Hooker
樟科 | Lauraceae

代码 (SpCode) = LITELO
个体数 (Individual number/20 hm^2) = 10383
最大胸径 (Max DBH) = 39.9 cm
重要值排序 (Importance value rank) = 2

常绿乔木，高达12 m；树皮灰黄色或褐色。小枝黄褐至灰褐色，密被褐色绒毛。叶互生，长圆形、长圆状披针形至倒披针形，革质，上面无毛，下面被短柔毛，沿中脉及侧脉有长柔毛，羽状脉。果长圆形,成熟时黑紫色。花期5～11月，果期翌年2～6月。

Evergreen trees, up to 12 m tall. Bark grayish yellow or brown. Branchlet yellowish brown to grayish brown, densely brown tomentose. Leaves alternate, oblong, elliptic-lanceolate or oblanceolate, leathery, glabrous adaxially, puberulent abaxially, viens and lateral veins long pubescence, pinninerved. Fruits oblong, black-purple at maturity. Fl. May - Nov., fr. Feb. - Jun. of following year.

树干　Trunk
摄影：杨庆松　Photo by: Yang Qingsong

枝叶　Branch and leaves
摄影：杨庆松　Photo by: Yang Qingsong

果枝　Fruiting branch
摄影：杨庆松　Photo by: Yang Qingsong

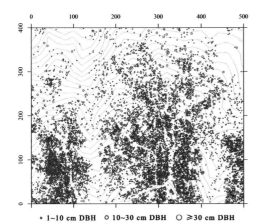

个体分布图　Distribution of individuals

径级分布表 DBH class

胸径等级 (Diameter class) (cm)	个体数 (No. of individuals in the plot)	比例 (Proportion) (%)
1～2	2643	25.46
2～5	4347	41.87
5～10	2173	20.93
10～20	1138	10.96
20～30	73	0.70
30～60	9	0.09
≥60	0	0.00

45 薄叶润楠 (华东楠)

Báo yè rùn nán | Thinleaf Machilus

Machilus leptophylla Hand.-Mazz.
樟科 | Lauraceae

代码 (SpCode) = MACLEP
个体数 (Individual number/20 hm^2) = 1163
最大胸径 (Max DBH) = 42.0 cm
重要值排序 (Importance value rank) = 17

常绿乔木，高达28 m；树皮灰褐色，平滑不裂。顶芽近球形，径可达2 cm。叶互生或在当年生枝上轮生，倒卵状长圆形，长14~24(32) cm，宽3.5~7(8) cm，先端短渐尖，基部楔形，坚纸质。圆锥花序6~10个，聚生嫩枝的基部。果球形，直径约1 cm。花期4月，果期7月。

Evergreen trees, up to 28 m tall. Bark grayish brown, smooth and no crack. Terminal buds subglobose, 2 cm in diam. Leaves alternate or verticillate on current-year-grown branches, obvate-elliptic, 14~24(~32) × 3.5~7(~8) cm, apex shortly acuminate, base cuneate, thinly papery. Panicles 6-10, congested on base of young branchlet. Fruit globose, 1 cm in diam. Fl. Apr., fr. Jul..

树干 Trunk
摄影：杨庆松 Photo by: Yang Qingsong

花枝 Flowering branch
摄影：杨庆松 Photo by: Yang Qingsong

果 Fruits
摄影：杨庆松 Photo by: Yang Qingsong

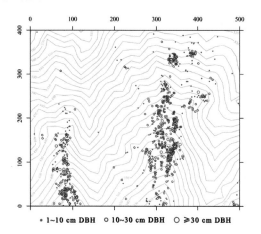

个体分布图 Distribution of individuals

径级分布表 DBH class

胸径区间 (Diameter class) (cm)	个体数 (No. of individuals in the plot)	比例 (Proportion) (%)
1~2	365	31.38
2~5	268	23.04
5~10	105	9.03
10~20	283	24.33
20~30	113	9.72
30~60	29	2.49
≥60	0	0.00

46 红楠　　　　　　　　　　　　　　　　　　　　　　　　Hóng nán | Red Nanmu

Machilus thunbergii Siebold & Zuccarini
樟科 | Lauraceae

代码 (SpCode) = MACTHU
个体数 (Individual number/20 hm^2) = 2802
最大胸径 (Max DBH) = 46.8 cm
重要值排序 (Importance value rank) = 10

常绿乔木，高20 m；树皮黄褐色。枝条多而伸展，紫褐色，老枝粗糙，嫩枝紫红色。叶倒卵形至倒卵状披针形，有刺激性气味；叶柄长1～3.5 cm；上面有浅槽，红色。花序顶生或在新枝上腋生，多花。果扁球形，直径8～10 mm，初时绿色，后变黑紫色。花期4月，果期6～7月。

Evergreen trees, up to 20 m tall. Bark yellowish brown. Branch multitudinous and extended, purple-brown, older branchlets rough, young branchlets purple-red. Leaves blade obovate or obovate-lanceolate, pungent smell. Petiole 1~3.5 cm, with shallow slot, red. Inflorescences terminal or axillary in new branchlets, many flowered. Fruti compressed globose, 8~10 mm in diam, green becoming dark purple. Fl. Apr., fr. Jun. - Jul..

树干　　Trunk
摄影：杨庆松　Photo by: Yang Qingsong

枝叶　　Branch and leaves
摄影：杨庆松　Photo by: Yang Qingsong

花枝　　Flowering branch
摄影：杨庆松　Photo by: Yang Qingsong

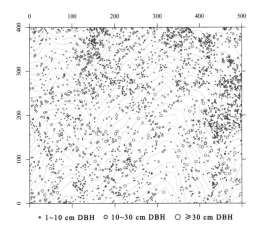

个体分布图　Distribution of individuals

径级分布表　DBH class

胸径等级 (Diameter class) (cm)	个体数 (No. of individuals in the plot)	比例 (Proportion) (%)
1～2	865	30.87
2～5	1028	36.69
5～10	439	15.67
10～20	323	11.53
20～30	76	2.71
30～60	71	2.53
≥60	0	0.00

47 浙江新木姜子

Zhè jiāng xīn mù jiāng zǐ | Zhejiang Newlitsea

Neolitsea aurata var. *chekiangensis* (Nakai) Yang et P. H. Huang
樟科 | Lauraceae

代码 (SpCode) = NEOAUR
个体数 (Individual number/20 hm^2) = 3203
最大胸径 (Max DBH) = 45.7 cm
重要值排序 (Importance value rank) = 12

常绿乔木，高达14 m；树皮灰褐色，平滑不裂。幼枝黄褐或红褐色，有锈色短柔毛。叶互生或聚生枝顶呈轮生状，披针形或倒披针形，革质，离基三出脉。伞形花序3～5个簇生于枝顶或节间。果椭圆形。花期3～4月，果期10～11月。

Evergreen trees, up to 14 m tall. Bark gray-brown, smooth and no crack. Branchlets yellow-brown or red-brown when young, rusty tomentose. Leaves alternate or clustered verticillately at apex of branchlet, lanceolate or oblanceolate, leathery, triplinerved. Umbel 3~5 cluster at apex of branchlet or internode. Fruit elliptic. Fl. Mar. - Apr., fr. Oct. - Nov..

幼苗　Seedling
摄影：刘何铭　Photo by: Liu Heming

枝叶　Branch and leaves
摄影：杨庆松　Photo by: Yang Qingsong

果枝　Fruiting branch
摄影：王希华　Photo by: Wang Xihua

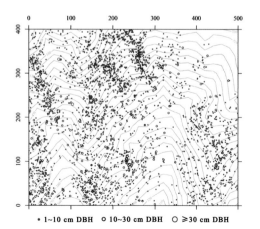

个体分布图　Distribution of individuals

径级分布表 DBH class

胸径区间 (Diameter class) (cm)	个体数 (No. of individuals in the plot)	比例 (Proportion) (%)
1～2	688	21.48
2～5	1366	42.65
5～10	934	29.16
10～20	204	6.37
20～30	7	0.22
30～60	4	0.12
≥60	0	0.00

48 紫楠

Zǐ nán | Purple Nanmu

Phoebe sheareri (Hemsl.) Gamble
樟科 | Lauraceae

代码 (SpCode) = PHOSHE
个体数 (Individual number/20 hm^2) = 180
最大胸径 (Max DBH) = 22.3 cm
重要值排序 (Importance value rank) = 53

常绿乔木，高20 m。小枝、叶柄及花序密被黄褐色或灰黑色柔毛或绒毛。叶革质，倒卵形、椭圆状倒卵形或阔倒披针形，长8～27 cm，宽4～9 cm，上面完全无毛或沿脉上有毛，下面密被黄褐色长柔毛。圆锥花序长7～15(18) cm，在顶端分枝。果卵形。花期4～5月，果期9～10月。

Evergreen trees, up to 20 m tall. Branchlets, petioles and inflorescences densely yellowish brown or gray-blackish pubescent to tomentose. Leaves leathery, obovate, elliptic-obovate or broadly oblanceolate, 8~27 × 4~9 cm, adaxially glabrous or hairy along veins, abaxially densely yellowish-brown pubescent. Panicles 7~15(~18) cm, branched at top of peduncle. Fruit ovoid. Fl. Apr. - May, fr. Sep. - Oct..

果 Fruits
摄影：杨庆松 Photo by: Yang Qingsong

枝叶 Branch and leaves
摄影：杨庆松 Photo by: Yang Qingsong

花序 Inflorescence
摄影：杨庆松 Photo by: Yang Qingsong

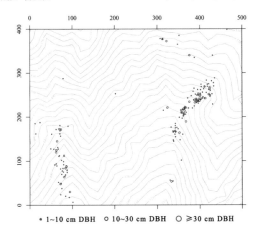

个体分布图 Distribution of individuals

径级分布表 DBH class

胸径等级 (Diameter class) (cm)	个体数 (No. of individuals in the plot)	比例 (Proportion) (%)
1～2	34	18.89
2～5	55	30.56
5～10	47	26.11
10～20	43	23.89
20～30	1	0.56
30～60	0	0.00
≥60	0	0.00

49 檫木　　　　　　　　　　　　　　　　　　　　　　　Chá mù | Sassafras

Sassafras tzumu (Hemsl.) Hemsl.
樟科 | Lauraceae

代码 (SpCode) = SASTZU

个体数 (Individual number/20 hm^2) = 267

最大胸径 (Max DBH) = 63.5 cm

重要值排序 (Importance value rank) = 20

落叶乔木，高可达35 m；树皮幼时黄绿色，平滑，老时变灰褐色，呈不规则纵裂。叶互生，聚集于枝顶，卵形或倒卵形，基部楔形，全缘或2~3浅裂，羽状脉或离基三出脉。花序顶生，先叶开放。果近球形。花期3~4月，果期5~9月。

Deciduous trees, up to 35 m tall. Bark yellow-green when young but gray-brown when mature, smooth, irregularly longitudinally fissured. Leaves alternate, clustered at apex of branchlet, ovate or obovate, base cuneate, margin or 2~3 lobed, pinninerved or triplinerved. Inflorescences terminal, appearing before leaves. Fruit subglobose. Fl. Mar. - Apr., fr. May - Sep..

树干　　Trunk
摄影：杨庆松　　Photo by: Yang Qingsong

叶　　Leaves
摄影：杨庆松　　Photo by: Yang Qingsong

花枝　　Flowering branches
摄影：杨庆松　　Photo by: Yang Qingsong

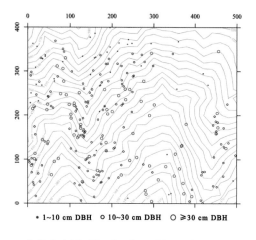

个体分布图 Distribution of individuals

径级分布表 DBH class

胸径区间 (Diameter class) (cm)	个体数 (No. of individuals in the plot)	比例 (Proportion) (%)
1~2	38	14.23
2~5	9	3.37
5~10	2	0.75
10~20	55	20.60
20~30	91	34.08
30~60	70	26.22
≥60	2	0.75

50 海金子 (崖花海桐)

Hǎi jīn zi | Anisetree-like Seatung

Pittosporum illicioides Makino
海桐花科 | Pittosporaceae

代码 (SpCode) = PITILL
个体数 (Individual number/20 hm^2) = 4
最大胸径 (Max DBH) = 2.1 cm
重要值排序 (Importance value rank) = 130

常绿灌木，高达5 m。叶生于枝顶，3~8片簇生呈假轮生状，薄革质，倒卵状披针形或倒披针形，5~10 cm，宽2.5~4.5 cm，基部窄楔形，常向下延，先端渐尖，下面浅绿色，上面深绿色，干后仍发亮。伞形花序顶生，花梗纤细，常向下弯。蒴果近圆形。花期4~5月，果期6~10月。

Evergreen shrubs, up to 5 m tall. Leaves 3~8-clustered at branchlet apex, appearing pseudoverticillate; thinly leathery, glabrous; obovate-lanceolate to oblanceolate, 5~10 × 2.5~4.5 cm; base narrowly cuneate, usually decurrent, apex obtuse or acuminate, leaf blade pale green abaxially, dark green adaxially, shiny after drying. Inflorescences terminal, umbellate, pedicel usually curved, slender. Capsule subglobose. Fl. Apr. - May, fr. Jun. - Oct..

叶　　Leaves
摄影：王樟华　Photo by: Wang Zhanghua

花　　Flowers
摄影：汪远　Photo by: Wang Yuan

果枝　　Fruiting branches
摄影：杨庆松　Photo by: Yang Qingsong

径级分布表 DBH class

胸径等级 (Diameter class) (cm)	个体数 (No. of individuals in the plot)	比例 (Proportion) (%)
1~2	3	75.00
2~5	1	25.00
5~10	0	0.00
10~20	0	0.00
20~30	0	0.00
30~60	0	0.00
≥60	0	0.00

个体分布图 Distribution of individuals

51 杨梅叶蚊母树

Yáng méi yè wén mǔ shù | Myrica-like Mosquitomam

Distylium myricoides Hemsl.
金缕梅科 | Hamamelidaceae

代码 (SpCode) = DISMYR

个体数 (Individual number/20 hm^2) = 6221

最大胸径 (Max DBH) = 37.8 cm

重要值排序 (Importance value rank) = 4

常绿乔木，高达10 m。嫩枝有鳞垢，老枝无毛，有皮孔，干后灰褐色。叶革质，矩圆形或倒披针形，长5~11 cm，宽2~4 cm，基部楔形，边缘全缘或上半部有锯齿，先端锐尖；叶柄长5~8 mm，有鳞垢。蒴果卵圆形，长1~1.2 cm，有黄褐色星毛。种子长6~7 mm。花期4~6月，果期7~8月。

Evergreen trees, up to 10 m tall. Young branches stellately lepidote, older growth drying gray-brown. Leaf leather, blade oblong or oblanceolate, 5~11 × 2~4 cm, base cuneate, margin entire or toothed above middle, apex acute. Petiole 5~8 mm, lepidote hairy. Capsules ovoid, 1~1.2 cm, stellately pubescent with yellow-brown hairs. Seeds 6~7 mm. Fl. Apr. -Jun., fr. Jul.-Aug..

幼苗　Seedling
摄影：刘何铭　Photo by: Liu Heming

枝叶　Branch and leaves
摄影：杨庆松　Photo by: Yang Qingsong

果枝　Fruiting branches
摄影：杨庆松　Photo by: Yang Qingsong

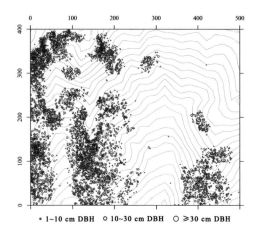

个体分布图　Distribution of individuals

径级分布表　DBH class

胸径区间 (Diameter class) (cm)	个体数 (No. of individuals in the plot)	比例 (Proportion) (%)
1~2	1698	27.29
2~5	1842	29.61
5~10	1246	20.03
10~20	1226	19.71
20~30	199	3.20
30~60	10	0.16
≥60	0	0.00

52 牛鼻栓

Niú bí shuān | China Fortunearia

Fortunearia sinensis Rehder et E. H. Wilson
金缕梅科 | Hamamelidaceae

代码 (SpCode) = FORSIN
个体数 (Individual number/20 hm^2) = 1
最大胸径 (Max DBH) = 6.9 cm
重要值排序 (Importance value rank) = 145

落叶灌木或小乔木，高5 m。叶倒卵形或倒卵状椭圆形，长7~16 cm，宽4~10 cm，膜质，先端锐尖，基部圆形或钝，稍偏斜，叶背面具长柔毛，沿中脉被短柔毛；边缘有锯齿；叶柄长4~10 mm。种子卵圆形，长约1 cm。花期4月，果期7~9月。

Deciduous shrubs or small trees, to 5 m tall. Leaf blade obovate or obovate-elliptic, 7~16 × 4~10 cm, membranous, apex acute, base rounded or obtuse, base oblique, abaxially villous, adaxially pubescent along midrib, margin dentate, petiole 4~10 mm. Seeds ovoid, 1 cm. Fl. Apr., fr. Jul. - Sep..

树干　　　Trunk
摄影：杨庆松　　Photo by: Yang Qingsong

果枝　　　Fruiting branches
摄影：汪远　　Photo by: Wang Yuan

叶　　　Leaves
摄影：杨庆松　　Photo by: Yang Qingsong

径级分布表 DBH class

胸径等级 (Diameter class) (cm)	个体数 (No. of individuals in the plot)	比例 (Proportion) (%)
1~2	0	0.00
2~5	0	0.00
5~10	1	100.00
10~20	0	0.00
20~30	0	0.00
30~60	0	0.00
≥60	0	0.00

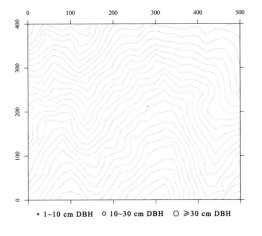

个体分布图 Distribution of individuals

53 枫香树

Fēng xiāng shù | Beautiful Sweetgum

Liquidambar formosana Hance
金缕梅科 | Hamamelidaceae

代码 (SpCode) = LIQFOR
个体数 (Individual number/20 hm^2) = 183
最大胸径 (Max DBH) = 81.5 cm
重要值排序 (Importance value rank) = 18

落叶乔木，高达30 m，树皮灰褐色。叶薄革质，阔卵形，掌状浅3裂，中央裂片较长，先端尾状渐尖，基部圆形；边缘有锯齿，齿尖有腺状突；叶柄长达11 cm。雄花短穗状花序常多个排成总状；雌花头状花序。种子多数，褐色，多角形或有窄翅。花期4～5月，果期7～10月。

Deciduous trees, up to 30 m tall, bark gray-brown. Leaf thinly leathery, blade broadly ovate, palmately 3-lobed and 3-veined, central lobe longer, apex caudate-acuminate, base rounded; margin glandular serrate; petiole up to 11 cm. Male inflorescence a short spike, several arranged in a raceme. Female inflorescence capitulum; Seeds many, brown. Fl. Apr. - May, fr. Jul. - Oct..

果枝　Fruiting branches
摄影：杨庆松　Photo by: Yang Qingsong

雌花序　Female inflorescences
摄影：杨庆松　Photo by: Yang Qingsong

雄花序　Male inflorescence
摄影：杨庆松　Photo by: Yang Qingsong

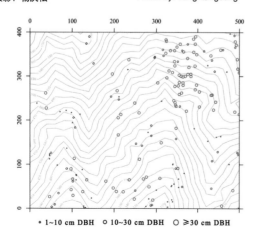
个体分布图　Distribution of individuals

径级分布表　DBH class

胸径区间 (Diameter class) (cm)	个体数 (No. of individuals in the plot)	比例 (Proportion) (%)
1～2	6	3.28
2～5	18	9.84
5～10	24	13.11
10～20	23	12.57
20～30	19	10.38
30～60	82	44.81
≥60	11	6.01

54 檵木 | Jì mù | China Loropetal

Loropetalum chinense (R. Br.) Oliver
金缕梅科 | Hamamelidaceae

代码 (SpCode) = LORCHI
个体数 (Individual number/20 hm^2) = 55
最大胸径 (Max DBH) = 18.3 cm
重要值排序 (Importance value rank) = 90

常绿灌木，有时为小乔木。叶革质，卵形，先端尖锐，基部钝，多少偏斜，背面密被星状短柔毛，正面疏生短柔毛或幼时具短柔毛，后脱落。花3~8朵簇生，有短花梗，白色，比新叶先开放，或与嫩叶同时开放。蒴果卵圆形，被褐色星状绒毛。花期3~5月，果期6~8月。

Evergreen shrubs, or small trees sometimes. Leaf leathery, blade ovate, apex acute, base asymmetrical, abaxially densely stellately pubescent, adaxially sparsely pubescent or stellately pubescent when young, glabrescent. Flowers 3~8 tufted, with short pedicel, white, open earlier than new leaves, or open with young leaves at the same time. Persistent floral cup 2/3~4/5 as long as capsule; Capsules ovoid, stellately tomentose, hairs brown. Fl. Mar. - May, fr. Jun. - Aug..

叶　Leaves
摄影：杨庆松　Photo by: Yang Qingsong

花枝　Flowering branch
摄影：杨庆松　Photo by: Yang Qingsong

果枝　Fruiting branches
摄影：杨庆松　Photo by: Yang Qingsong

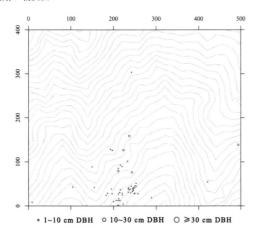
个体分布图　Distribution of individuals

径级分布表　DBH class

胸径等级 (Diameter class) (cm)	个体数 (No. of individuals in the plot)	比例 (Proportion) (%)
1~2	4	7.27
2~5	31	56.36
5~10	14	25.45
10~20	6	10.91
20~30	0	0.00
30~60	0	0.00
≥60	0	0.00

55 迎春樱桃

Yíng chūn yīng táo | Discoid Cherry

Cerasus discoidea T. T. Yu et C. L. Li
蔷薇科 | Rosaceae

代码 (SpCode) = CERDIS
个体数 (Individual number/20 hm^2) = 77
最大胸径 (Max DBH) = 15.2 cm
重要值排序 (Importance value rank) = 78

落叶小乔木，高5～8 m。小枝紫褐色，嫩枝被疏柔毛，后脱落无毛。叶片倒卵状长圆形或长椭圆形，长4～8 cm，宽1.5～3.5 cm，先端骤尾尖或尾尖，基部楔形，稀近圆形，边有缺刻状尖锯齿；叶柄长5～7 mm，顶端有1～3腺体。伞形花序。核果成熟时红色。花期3月，果期5月。

Deciduous small trees, 5~8 m tall. Branchlets purplish brown; young branchlets pilose, glabrescent. Leaf blade obovate-oblong to elliptic, 4~8 × 1.5~3.5 cm, base cuneate to rarely subrounded, margin shallowly obtusely serrulate and teeth with a minute conical apical gland, apex caudate, cauda acutely incised serrate. Petiole 5~7 mm, apex with 1~3 nectaries; Inflorescences umbellate. Drupe red. Fl. Mar., fr. May.

树干　Trunk
摄影：杨庆松　Photo by: Yang Qingsong

枝叶　Branch and leaves
摄影：杨庆松　Photo by: Yang Qingsong

果枝　Fruiting branch
摄影：杨庆松　Photo by: Yang Qingsong

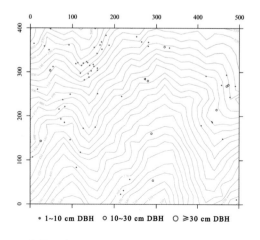

个体分布图　Distribution of individuals

径级分布表　DBH class

胸径区间 (Diameter class) (cm)	个体数 (No. of individuals in the plot)	比例 (Proportion) (%)
1～2	11	14.29
2～5	11	14.29
5～10	46	59.74
10～20	9	11.69
20～30	0	0.00
30～60	0	0.00
≥60	0	0.00

56 大叶早樱

Dà yè zǎo yīng | Higan Cherry

Cerasus subhirtella (Miq.) Sok.
蔷薇科 | Rosaceae

代码 (SpCode) = CERSUB
个体数 (Individual number/20 hm^2) = 30
最大胸径 (Max DBH) = 65.3 cm
重要值排序 (Importance value rank) = 46

落叶乔木，高10～20 m，树皮灰褐色。嫩枝绿色，密被白色短柔毛。冬芽卵形，鳞片先端有疏毛。叶片卵形至卵状长圆形，边有细锐锯齿和重锯齿，侧脉直出，几平行，有10～14对；叶柄长5～8 mm，被白色短柔毛。花序伞形，有花2～3朵，花叶同开。核果卵球形，黑色。花期4月，果期6月。

Deciduous trees 10~20 m tall. Bark grayish brown. Young branchlets green, densely white pubescent. Winter buds ovoid; bud scale margin pilose. Leaf blade ovate to ovate-oblong, base broadly cuneate, margin sharply biserrate; secondary veins 10~14 on each side of midvein, straight and parallel. Petiole 5~8 mm, white pubescent; Inflorescences umbellate, 2~3 flowered. Flowers opening at same time as leaves. Drupe black, ovoid. Fl. Apr., fr. Jun..

树干　Trunk
摄影：杨庆松　Photo by: Yang Qingsong

叶　Leaf
摄影：杨庆松　Photo by: Yang Qingsong

花　Flowers
摄影：汪远　Photo by: Wang Yuan

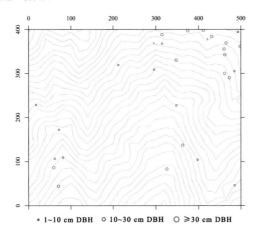

个体分布图　Distribution of individuals

径级分布表　DBH class

胸径等级 (Diameter class) (cm)	个体数 (No. of individuals in the plot)	比例 (Proportion) (%)
1～2	1	3.33
2～5	0	0.00
5～10	1	3.33
10～20	7	23.33
20～30	6	20.00
30～60	12	40.00
≥60	3	10.00

57 湖北山楂

Hú běi shān zhā | Hubei Hawthorn

Crataegus hupehensis Sarg.
蔷薇科 | Rosaceae

代码 (SpCode) = CRAHUP
个体数 (Individual number/20 hm^2) = 1
最大胸径 (Max DBH) = 12.0 cm
重要值排序 (Importance value rank) = 137

落叶小乔木或灌木，高3~5 m；枝具刺，也常无刺。叶片卵形至卵状长圆形，基部宽楔形或近圆形，边缘有圆钝锯齿，上半部具2~4对浅裂片，裂片卵形。托叶草质，披针形或镰刀形，边缘具腺齿，早落。伞房花序，直径3~4 cm；花瓣卵形，白色。果实深红色，有斑点。花期5~6月，果期8~9月。

Deciduous small trees or shrubs, 3~5 m tall; branches sparsely thorny, sometimes unarmed. Leaf blade ovate or ovate-oblong, base broadly cuneate or subrounded, margin crenate-serrate, with 2~4 pairs of shallow lobes at apical part; lobes ovate. Stipules caducous, lanceolate or falcate, herbaceous, margin glandular serrate, apex acuminate. Corymb 3~4 cm in diam. Petals white, ovate. Pome red, subglobose. Fl. May - Jun., fr. Aug. - Sep.

花枝　　Flowering branch
摄影：杨庆松　Photo by: Yang Qingsong

树干　　Trunks
摄影：王樟华　Photo by: Wang Zhanghua

果枝　　Fruiting branch
摄影：杨庆松　Photo by: Yang Qingsong

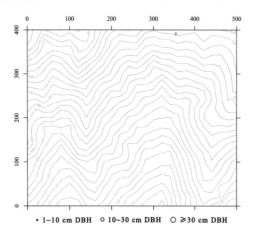
个体分布图 Distribution of individuals

径级分布表 DBH class

胸径区间 (Diameter class) (cm)	个体数 (No. of individuals in the plot)	比例 (Proportion) (%)
1~2	0	0.00
2~5	0	0.00
5~10	0	0.00
10~20	1	100.00
20~30	0	0.00
30~60	0	0.00
≥60	0	0.00

58 腺叶桂樱

Xiàn yè guì yīng | Glandleaf Cherrylaurel

Laurocerasus phaeosticta (Hance) Schneid.
蔷薇科 | Rosaceae

代码 (SpCode) = LAUPHA
个体数 (Individual number/20 hm^2) = 679
最大胸径 (Max DBH) = 15.5 cm
重要值排序 (Importance value rank) = 37

常绿灌木或小乔木，高4~12 m。小枝暗紫褐色。叶片革质，两面无毛，叶背面散生黑色斑点，基部近叶缘常有2枚较大扁平基腺，叶边全缘或不育小枝有尖锐的锯齿，先端长尾状。叶柄长4~8 mm，无毛。总状花序单生于叶腋，花数朵至10朵，花瓣白色，近圆形。果实近球形或横向椭圆形，直径8~10 mm，紫黑色，无毛。花期4~5月，果期7~10月。

Evergreen shrubs or small trees, 4~12 m tall. Branchlets dark purplish brown. Leaf blade leathery, both surfaces glabrous, abaxially scattered black punctate, base cuneate and with 2 large flat nectaries near margin, margin entire or on sterile branchlets acutely serrate, apex long caudate. Petiole 4~8 mm, glabrous. Raceme solitary in distal leaf axil, flowers up to 10. Petals white, round. Drupe purplish black, subglobose to transversely ellipsoid, 8~10 mm in diam., glabrous. Fl. Apr. - May, fr. Jul. - Oct..

枝叶　Branch and leaves
摄影：杨庆松　Photo by: Yang Qingsong

花枝　Flowering branch
摄影：杨庆松　Photo by: Yang Qingsong

果枝　Fruiting branches
摄影：杨庆松　Photo by: Yang Qingsong

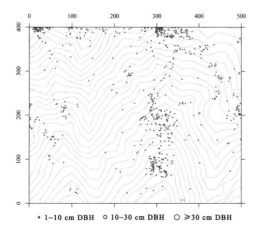

个体分布图　Distribution of individuals

径级分布表 DBH class

胸径等级 (Diameter class) (cm)	个体数 (No. of individuals in the plot)	比例 (Proportion) (%)
1~2	199	29.31
2~5	329	48.45
5~10	132	19.44
10~20	19	2.80
20~30	0	0.00
30~60	0	0.00
≥60	0	0.00

59 刺叶桂樱

Cì yè guì yīng | Spinyleaf Cherrylaurel

Laurocerasus spinulosa (Sieb. et Zucc.) Schneid.
蔷薇科 | Rosaceae

代码 (SpCode) = LAUSPI
个体数 (Individual number/20 hm^2) = 43
最大胸径 (Max DBH) = 37.4 cm
重要值排序 (Importance value rank) = 74

常绿乔木，高可达20 m，稀为灌木。小枝紫褐色或黑褐色，无毛，具明显皮孔。叶片草质至薄革质，长圆形或倒卵状长圆形，先端渐尖至尾尖，基部宽楔形至近圆形，边缘不平而常呈波状，中部以上或近顶端常具少数针状锐锯齿，两面无毛。总状花序生于叶腋，单生；花瓣圆形，白色。果实椭圆形，褐色至黑褐色。花期9~10月，果期11月至翌年3月。

Evergreen trees to 20 m tall, rarely shrubs. Branchlets purplish brown to blackish brown, glabrous, prominently lenticellate. Leaf blade oblong to obovate-oblong, herbaceous to thinly leathery, both surfaces glabrous, base broadly cuneate to subrounded and often unequal, margin undulate and with a few acicular teeth apically from middle to near apex, apex acuminate to caudate. Racemes axillary, solitary. Petals white, orbicular. Drupe brown to blackish brown, ellipsoid. Fl. Sep. - Oct., fr. Nov. - Mar. of following year.

枝叶　Branch and leaves
摄影：杨庆松　Photo by: Yang Qingsong

叶　Leaf
摄影：杨庆松　Photo by: Yang Qingsong

果枝　Fruiting branch
摄影：杨庆松　Photo by: Yang Qingsong

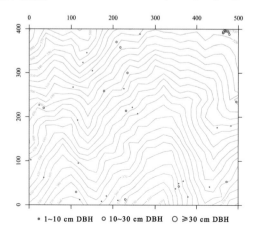
个体分布图　Distribution of individuals

径级分布表　DBH class

胸径区间 (Diameter class) (cm)	个体数 (No. of individuals in the plot)	比例 (Proportion) (%)
1~2	5	11.63
2~5	11	25.58
5~10	8	18.60
10~20	10	23.26
20~30	8	18.60
30~60	1	2.33
≥60	0	0.00

60 中华石楠

Zhōng huá shí nán | China Photinia

Photinia beauverdiana Schneid.
蔷薇科 | Rosaceae

代码 (SpCode) = PHOBEA
个体数 (Individual number/20 hm^2) = 1
最大胸径 (Max DBH) = 3.3 cm
重要值排序 (Importance value rank) = 148

落叶灌木或小乔木，高3～10 m。小枝紫褐色，散生灰色皮孔，通常无毛，具散生灰色皮孔。叶片薄纸质，长圆形、倒卵状长圆形或卵状披针形，上面光亮，基部圆形或楔形，边缘疏生具腺锯齿，先端锐尖；叶柄长5～10 mm，微被柔毛。复伞花序，花多数；花瓣白色。果实紫红色，卵形，长7～8 mm，直径5～6 mm。花期5月，果期7～8月。

Deciduous shrubs or small trees, 3~10 m tall. Branchlets purplish brown, usually glabrous, with scattered gray lenticels. Leaf blade oblong, ovate, or elliptic to obovate, glabrous, base cuneate to rounded, margin serrate, apex acute; petiole 5~10 mm, puberulous. Compound corymbs terminal, many flowered; petals white. Fruit purplish red, ovoid or subglobose, 7~8 × 5~6 mm. Fl. May, fr. Jul. - Aug..

树干　Trunk
摄影：杨庆松　Photo by: Yang Qingsong

枝叶　Branch and leaves
摄影：杨庆松　Photo by: Yang Qingsong

果　Fruits
摄影：杨庆松　Photo by: Yang Qingsong

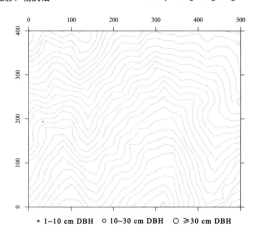

个体分布图 Distribution of individuals

径级分布表 DBH class

胸径等级 (Diameter class) (cm)	个体数 (No. of individuals in the plot)	比例 (Proportion) (%)
1～2	0	0.00
2～5	1	100.00
5～10	0	0.00
10～20	0	0.00
20～30	0	0.00
30～60	0	0.00
≥60	0	0.00

61 光叶石楠

Guāng yè shí nán | Japan Photinia

Photinia glabra (Thunb.) Maxim.
蔷薇科 | Rosaceae

代码 (SpCode) = PHOGLA
个体数 (Individual number/20 hm^2) = 680
最大胸径 (Max DBH) = 30.7 cm
重要值排序 (Importance value rank) = 34

常绿小乔木，高可达7 m。叶片革质，幼时呈红色，椭圆形、长圆形或长圆倒卵形，长5~9 cm，宽2~4 cm，先端渐尖，基部楔形，边缘有疏生浅钝细锯齿，两面无毛；叶柄长1~1.5 cm，无毛。顶生复伞房花序，花多数；花瓣白色。果实卵形，红色。花期4~5月，果期9~10月。

Evergreen trees, up to 7 m tall. Leaf blade initially reddish, elliptic, oblong, or oblong-obovate, 5~9 × 2~4 cm, both surfaces glabrous, apex acuminate, base cuneate, margin with sparse, shallowly crenulate teeth; petiole 1~1.5 cm, glabrous. Compound corymbs terminal, numerous flowered; petals white, obovate. Fruit red, obovate or ovoid. Fl. Apr. - May, fr. Sep. - Oct..

树干　　Trunk
摄影：杨庆松　Photo by: Yang Qingsong

花枝　　Flowering branch
摄影：杨庆松　Photo by: Yang Qingsong

果枝　　Fruiting branch
摄影：杨庆松　Photo by: Yang Qingsong

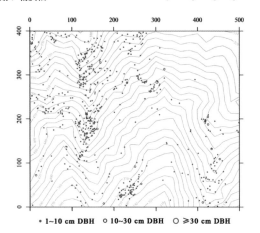

个体分布图 Distribution of individuals

径级分布表 DBH class

胸径区间 (Diameter class) (cm)	个体数 (No. of individuals in the plot)	比例 (Proportion) (%)
1~2	83	12.21
2~5	345	50.74
5~10	234	34.41
10~20	17	2.50
20~30	0	0.00
30~60	1	0.15
≥60	0	0.00

62 小叶石楠

Xiǎo yè shí nán | Littleleaf Photinia

Photinia parvifolia (Pritz.) Schneid.
蔷薇科 | Rosaceae

代码 (SpCode) = PHOPAR
个体数 (Individual number/20 hm^2) = 11
最大胸径 (Max DBH) = 11.4 cm
重要值排序 (Importance value rank) = 114

落叶灌木，高1~3 m。枝纤细，小枝红褐色，初时具柔毛，后脱落，有黄色散生皮孔。叶片草质，椭圆形、椭圆卵形或菱状卵形，先端渐尖或尾尖，基部宽楔形或近圆形，边缘有具腺尖锐锯齿；叶柄长1~2 mm。花2~9朵，成伞形花序。果实椭圆形或卵形，种子卵形；果梗长1~2.5 cm，密布疣点。花期4~5月，果期7~8月。

Deciduous shrubs, 1~3 m tall. Branchlets reddish brown, slender, pilose when young, glabrous, with scattered yellow lenticels. Petiole 1~2 mm; leaf blade ovate, elliptic, or elliptic-ovate to rhombic-ovate, base cuneate to subrounded, margin sharply serrate, apex acuminate or caudate. Inflorescences umbellate, 2~9 floweres. Fruit ellipsoid or ovoid, seeds ovoid; fruiting pedicels 1~2.5 cm, with dense lenticels. Fl. Apr. - May, fr. Jul. - Aug..

果　Fruits
摄影：杨庆松　Photo by: Yang Qingsong

花　Flowers
摄影：严靖　Photo by: Yan Jing

叶　Leaf
摄影：杨庆松　Photo by: Yang Qingsong

径级分布表 DBH class

胸径等级 (Diameter class) (cm)	个体数 (No. of individuals in the plot)	比例 (Proportion) (%)
1~2	2	18.18
2~5	6	54.55
5~10	2	18.18
10~20	1	9.09
20~30	0	0.00
30~60	0	0.00
≥60	0	0.00

个体分布图 Distribution of individuals

63 石斑木　　　Shí bān mù | Hongkong Raphiolepis

Rhaphiolepis indica (L.) Lindl.
蔷薇科 | Rosaceae

代码 (SpCode) = RHAIND

个体数 (Individual number/20 hm^2) = 66

最大胸径 (Max DBH) = 4.5 cm

重要值排序 (Importance value rank) = 93

常绿灌木，稀小乔木，高可达4 m。叶片集生于枝顶，卵形、长圆形，稀倒卵形或长圆披针形，长(2) 4~8 cm，宽1.5~4 cm，先端圆钝、急尖、渐尖或长尾尖，边缘具细钝锯齿，上面光亮，平滑无毛，网脉不显明，下面色淡，无毛或被稀疏绒毛，叶脉稍凸起，网脉明显。顶生圆锥花序或总状花序。果实球形，紫黑色。花期4月，果期7~8月。

Evergreen shrubs, rarely small trees, to 4 m tall. Leaf blade ovate, oblong, rarely obovate or oblong-lanceolate, (2~) 4~8 × 1.5~4 cm, leathery, veins prominent abaxially, reticulate veins conspicuous abaxially and conspicuous or not adaxially, abaxially pale, glabrous or sparsely tomentose, adaxially lustrous, glabrous, base attenuate, margin crenulate, serrate, or obtusely serrate, apex obtuse, acute, acuminate, or long caudate. Panicle or racemes terminal. Pome purplish black, globose. Fl. Apr., fr. Jul. - Aug..

树干　Trunk
摄影：杨庆松　Photo by: Yang Qingsong

花枝　Flowering branch
摄影：杨庆松　Photo by: Yang Qingsong

果枝　Fruiting branches
摄影：杨庆松　Photo by: Yang Qingsong

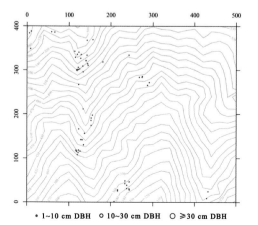
个体分布图　Distribution of individuals

径级分布表　DBH class

胸径区间 (Diameter class) (cm)	个体数 (No. of individuals in the plot)	比例 (Proportion) (%)
1~2	38	57.58
2~5	28	42.42
5~10	0	0.00
10~20	0	0.00
20~30	0	0.00
30~60	0	0.00
≥60	0	0.00

64 山槐（山合欢） Shān huái | Wild Siris

Albizia kalkora (Roxb.) Prain
豆科 | Fabaceae

代码 (SpCode) = ALBKAL
个体数 (Individual number/20 hm^2) = 3
最大胸径 (Max DBH) = 15.4 cm
重要值排序 (Importance value rank) = 119

落叶小乔木或灌木，高3~8 m。枝条暗褐色，被短柔毛，有显著皮孔。二回羽状复叶；羽片2~4对；小叶5~14对，两面均被短柔毛，中脉稍偏于上侧。头状花序2~7枚生于叶腋，或于枝顶排成圆锥花序。荚果带状，长7~17 cm，宽1.5~3 cm，深棕色，嫩荚密被短柔毛，老时无毛；种子4~12颗，倒卵形。花期5~6月，果期8~10月。

Deciduous small trees, or shrubs, 3~8 m tall. Branchlets dark brown, pubescent, with conspicuous lenticels. Bipinnately compound leaf, pinnae 2~4 pairs; leaflets 5~14 pairs, both surfaces pubescent, main vein slightly close to upper margin, base oblique, apex obtuse, mucronate. Heads 2~7, axillary or terminal, arranged in panicles. Legume dehiscent, ligulate, 7~17 × 1.5~3 cm, pubescent when young, glabrescent when mature. Seeds 4~12, obovoid or suborbicular; pleurogram oblong. Fl. May - Jun., fr. Aug. - Oct..

树干　Trunk
摄影：杨庆松　Photo by: Yang Qingsong

叶　Leaves
摄影：杨庆松　Photo by: Yang Qingsong

果枝　Fruiting branches
摄影：汪远　Photo by: Wang Yuan

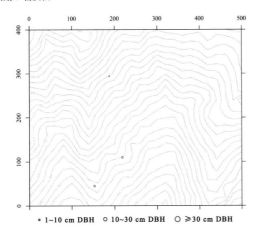
个体分布图　Distribution of individuals

径级分布表　DBH class

胸径等级 (Diameter class) (cm)	个体数 (No. of individuals in the plot)	比例 (Proportion) (%)
1~2	0	0.00
2~5	0	0.00
5~10	1	33.33
10~20	2	66.67
20~30	0	0.00
30~60	0	0.00
≥60	0	0.00

65 黄檀

Huáng tán | Hubei Rosewood

Dalbergia hupeana Hance
豆科 | Fabaceae

代码 (SpCode) = DALHUP

个体数 (Individual number/20 hm^2) = 192

最大胸径 (Max DBH) = 39.0 cm

重要值排序 (Importance value rank) = 35

落叶乔木，高10～20 m；树皮暗灰色，呈薄片状剥落。幼枝淡绿色，无毛。羽状复叶长15～25 cm；小叶7～11，近革质，椭圆形至长圆状椭圆形，先端钝或稍凹入，基部圆形或阔楔形、两面无毛，细脉隆起，上面有光泽。圆锥花序顶生或腋生，花冠白色或淡紫色。荚果长圆形或阔舌状，种子肾形。花期5～7月，果期8-9月。

Deciduous trees, 10~20 m tall. Bark dull gray, flaky exfoliating; young shoots pale green, glabrous. Leaves 15~25 cm; leaflets 7~11, elliptic to oblong-elliptic, subleathery, both surfaces glabrous, shiny adaxially, veinlets prominent, base rounded or broadly cuneate, apex obtuse or slightly emarginate. Panicles terminal or extending into axils of uppermost leaves, corolla white or light purple Seeds reniform. Fl. May - Jul., fr. Aug. - Sep..

树干　　Trunk
摄影：杨庆松　Photo by: Yang Qingsong

花枝　　Flowering branches
摄影：葛斌杰　Photo by: Ge Binjie

果枝　　Fruiting branch
摄影：汪远　Photo by: Wang Yuan

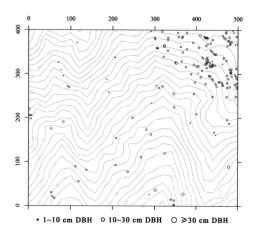

个体分布图　Distribution of individuals

径级分布表 DBH class

胸径区间 (Diameter class) (cm)	个体数 (No. of individuals in the plot)	比例 (Proportion) (%)
1～2	1	0.52
2～5	9	4.69
5～10	24	12.50
10～20	115	59.90
20～30	36	18.75
30～60	7	3.65
≥60	0	0.00

66 楝叶吴萸 (臭辣树)

liàn yè wú yú | Farges Evodia

Tetradium glabrifolium (Champ. ex Benth.) Hartley
芸香科 | Rutaceae

代码 (SpCode) = TETGLA
个体数 (Individual number/20 hm^2) = 4
最大胸径 (Max DBH) = 6.3 cm
重要值排序 (Importance value rank) = 127

落叶乔木，高达17 m。树皮平滑，暗灰色，嫩枝紫褐色，散生小皮孔。叶有小叶5～9 (11) 片，小叶倒披针形或椭圆形或椭圆状长圆形，网状细脉背面清晰、密集，在侧生小叶基部狭楔形到近圆形至近截形，在顶生小叶狭楔形至楔形，边缘全缘或具细圆齿，先端渐尖。花序顶生，花甚多。种子近球形至卵圆形到宽椭圆形，2.5～4 mm，褐黑色。花期6～8月，果期8～12月。

Deciduous trees, up to 17 m tall. Balk dull gray, smooth. Young branchlets purple brown, scattered lenticels. Leaves with 5~9 (11) leaflet; leaflet blades broadly ovate to lanceolate, reticulate veinlets abaxially clearly defined and dense, base in lateral leaflets narrowly cuneate to subrounded to subtruncate and in terminal leaflet narrowly cuneate to cuneate, margin entire or crenulate, apex acuminate. Inflorescence apical. Seeds subglobose to ovoid to broadly ellipsoid, 2.5~4 mm, blackish brown. Fl. Jun. - Aug., fr. Aug. - Dec..

树干 Trunk
摄影：杨庆松 Photo by: Yang Qingsong

枝叶 Branch and leaves
摄影：杨庆松 Photo by: Yang Qingsong

叶 Leaves
摄影：杨庆松 Photo by: Yang Qingsong

径级分布表 DBH class

胸径等级 (Diameter class) (cm)	个体数 (No. of individuals in the plot)	比例 (Proportion) (%)
1～2	1	25.00
2～5	2	50.00
5～10	1	25.00
10～20	0	0.00
20～30	0	0.00
30～60	0	0.00
≥60	0	0.00

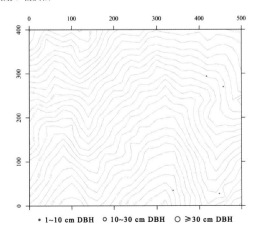

个体分布图 Distribution of individuals

67 密果吴萸 (吴茱萸)   Mì guǒ wú yú | Medicinal Evodia

Tetradium ruticarpum (A. Juss.) Hartley
芸香科 | Rutaceae

代码 (SpCode) = TETRUT
个体数 (Individual number/20 hm^2) = 1
最大胸径 (Max DBH) = 2.3 cm
重要值排序 (Importance value rank) = 151

落叶灌木或小乔木，高3～5 m。复叶有5～13小叶；小叶叶片椭圆形至卵、披针形、倒披针形或倒卵形，小叶较大，宽达7 cm，略厚纸质，两面密被长毛；花序顶生，花序轴被红褐色长毛；果序密集成卵球形蓇葖果近球形，每果种子1但与败育种子成对，种子褐黑色。花期4～6月，果期8～11月。

Deciduous shrubs or small trees, 3~5 m tall. Leaves with 5~13 foliolate; leaflet blades elliptic to ovate or sometimes lanceolate, oblanceolate, or obovate, base in lateral leaflets obtuse to narrowly cuneate or sometimes rounded to cuneate or rarely attenuate, margin entire or irregularly crenulate, apex acuminate. Follicles subglobose, seeds 1 per follicle but paired with an abortive seed, blackish brown. Fl. Apr. - Jun., fr. Aug. - Nov..

花序   Infloresceen
摄影：汪远   Photo by: Wang Yuan

枝叶   Branchs and leaves
摄影：葛斌杰   Photo by: Ge Binjie

果枝   Fruiting branch
摄影：汪远   Photo by: Wang Yuan

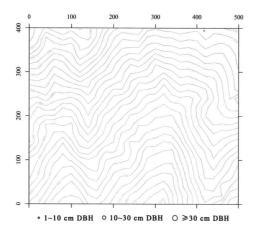

个体分布图 Distribution of individuals

径级分布表 DBH class

胸径区间 (Diameter class) (cm)	个体数 (No. of individuals in the plot)	比例 (Proportion) (%)
1～2	0	0.00
2～5	1	100.00
5～10	0	0.00
10～20	0	0.00
20～30	0	0.00
30～60	0	0.00
≥60	0	0.00

68 湖北算盘子

Hú běi suàn pán zi | Hubei Glochidion

Glochidion wilsonii Hutch.
大戟科 | Euphorbiaceae

代码 (SpCode) = GLOWIL
个体数 (Individual number/20 hm^2) = 1
最大胸径 (Max DBH) = 10.1cm
重要值排序 (Importance value rank) = 141

落叶灌木，高1～4 m；枝条具棱，灰褐色；小枝直而开展；除叶柄外，全株均无毛。叶片纸质，披针形或斜披针形，顶端短渐尖或急尖，基部钝或宽楔形，上面绿色，下面带灰白色；中脉两面凸起；叶柄长3～5 mm；托叶卵状披针形。花绿色，雌雄同株。蒴果扁球状，直径约1.5 cm。花期4～7月，果期6～9月。

Deciduous shrubs 1~4 m tall, branches angular, gray-brown; branchlets spreading, glabrous throughout except sometimes for pubescent petiole. Leaf blade lanceolate or obliquely lanceolate, papery, green adaxially, gray-white abaxially, base obtuse or broadly cuneate, apex acute or shortly acuminate; midvein elevated on both surfaces. Stipules ovate-lanceolate; petiole 3~5 mm. Flowers green, monoecious. Capsules depressed globose, ca. 1.5 cm in diam. Fl. Apr. - Jul., fr. Jun. - Sep..

幼苗　Seedling
摄影：杨庆松　Photo by: Yang Qingsong

枝叶　Branch and leaves
摄影：杨庆松　Photo by: Yang Qingsong

果枝　Fruiting branch
摄影：杨庆松　Photo by: Yang Qingsong

径级分布表　DBH class

胸径等级 (Diameter class) (cm)	个体数 (No. of individuals in the plot)	比例 (Proportion) (%)
1～2	0	0.00
2～5	0	0.00
5～10	0	0.00
10～20	1	100.00
20～30	0	0.00
30～60	0	0.00
≥60	0	0.00

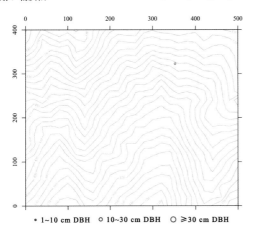
• 1~10 cm DBH　○ 10~30 cm DBH　○ ≥30 cm DBH

个体分布图　Distribution of individuals

69 白背叶

Bái bèi yè | Whitebackleaf

Mallotus apelta (Lour.) Muell.-Arg.
大戟科 | Euphorbiaceae

代码 (SpCode) = MALAPE
个体数 (Individual number/20 hm^2) = 179
最大胸径 (Max DBH) = 7.7 cm
重要值排序 (Importance value rank) = 70

落叶灌木或小乔木，高1～6 m。嫩枝被白色和褐色星状柔毛。叶柄5～15 cm，被白色柔毛；叶片宽卵形，纸质，背面带白色微柔毛和散生的橙黄色腺体，正面脱落无毛或疏生星状毛，基部截形或楔形，稀稍心形，具2基部腺体。花雌雄异株，为圆锥花序或穗状。雄花、雌花及果，均密被白色绒毛。种子卵球形，黑色。花期5～9月，果期8～11月。

Deciduous shrubs or small trees 1~6 m tall. Branchlets whitish and brownish stellate-tomentulose when young. Petiole 5~15 cm, whitish tomentulose; leaf blade broadly ovate, papery, abaxially whitish tomentulose and scattered orange glandular-scaly, adaxially glabrescent or sparsely stellate-pilosulose, base truncate or cuneate, rarely slightly cordate, with 2 basal glands. Dioecism, panicle or spike. Flowers and capsule, both whitish tomentulose. Seeds ovoid, black. Fl. May - Sep., fr. Aug. - Nov..

幼苗　Seedling
摄影：杨庆松　Photo by: Yang Qingsong

叶　Leaves
摄影：杨庆松　Photo by: Yang Qingsong

果枝　Fruiting branch
摄影：杨庆松　Photo by: Yang Qingsong

径级分布表 DBH class

胸径区间 (Diameter class) (cm)	个体数 (No. of individuals in the plot)	比例 (Proportion) (%)
1～2	97	54.19
2～5	79	44.13
5～10	3	1.68
10～20	0	0.00
20～30	0	0.00
30～60	0	0.00
≥60	0	0.00

个体分布图 Distribution of individuals

70 野桐   Yě tóng | Mallotus

Mallotus tenuifolius Pax
大戟科 | Euphorbiaceae

代码 (SpCode) = MALTEN
个体数 (Individual number/20 hm^2) = 6
最大胸径 (Max DBH) = 2.4 cm
重要值排序 (Importance value rank) = 125

落叶灌木，2～4 m高。嫩枝暗褐色，被星状微绒毛。叶柄长3～17 cm；叶片近圆形卵形或菱形卵形，有时波状，纸质，背面疏生星状小柔毛，疏生黄色腺鳞，正面无毛，基部楔形，钝，或稍心形，有时稍盾形，2 (或4) 腺体。蒴果疏生星状毛和腺鳞。种子近球形，褐色或黑色。花期4～6月，果期7～8月。

Deciduous shrubs 2~4 m tall. Branchlets dull brownish stellate-tomentulose when young. Petiole 3~17 cm; leaf blade suborbicular-ovate or rhombic-ovate, sometimes repand-tricuspidate, papery, abaxially sparsely stellate-pilosulose, sparsely yellowish glandular-scaly, adaxially glabrous, base cuneate, obtuse, or slightly cordate, sometimes slightly peltate, with 2(or 4) glands. Capsule with reddish orange glandular-scaly and sparsely softly spiny. Seeds subglobose, brown or black. Fl. Apr. - Jun., fr. Jul. - Aug..

叶背   Leaf abaxial surface
摄影：葛斌杰   Photo by: Ge Binjie

花枝   Flowering branch
摄影：杨庆松   Photo by: Yang Qingsong

果   Fruits
摄影：杨庆松   Photo by: Yang Qingsong

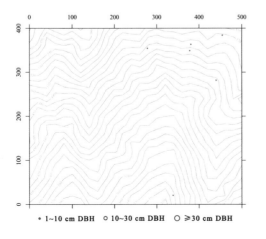

个体分布图 Distribution of individuals

径级分布表 DBH class

胸径等级 (Diameter class) (cm)	个体数 (No. of individuals in the plot)	比例 (Proportion) (%)
1～2	5	83.33
2～5	1	16.67
5～10	0	0.00
10～20	0	0.00
20～30	0	0.00
30～60	0	0.00
≥60	0	0.00

71 青灰叶下珠

Qīng huī yè xià zhū | Greyblue Underleaf Pearl

Phyllanthus glaucus Wall. ex Muell.-Arg.
大戟科 | Euphorbiaceae

代码 (SpCode) = PHYGLA
个体数 (Individual number/20 hm^2) = 3
最大胸径 (Max DBH) = 9.5 cm
重要值排序 (Importance value rank) = 129

落叶灌木，高达4 m；枝条圆柱形，小枝细柔；全株无毛。叶片膜质，椭圆形或长圆形，顶端急尖，有小尖头，基部钝至圆，下面稍苍白色；侧脉每边8～10条；叶柄长2～4 mm；托叶卵状披针形，膜质。花数朵簇生于叶腋。蒴果浆果状，直径约1 cm，成熟时紫黑色，基部有宿存的萼片；种子黄褐色。花期4～7月，果期7～10月。

Deciduous shrubs up to 4 m tall, monoecious, glabrous throughout; branches terete; branchlets delicate. Leaf blade elliptic or oblong, membranous, slightly glaucous abaxially, base obtuse to rounded, apex acute, apiculate; lateral veins 8~10 pairs; stipules ovate-lanceolate, membranous; petiole 2~4 mm. Inflorescence an axillary fascicle. Fruit a berry, ca. 1 cm in diam., black-purple. Seeds tawny. Fl. Apr. - Jul., fr. Jul. - Oct..

叶　Leaves
摄影：葛斌杰　Photo by: Ge Binjie

花枝　Flowering branches
摄影：胡瑾瑾　Photo by: Hu Jinjin

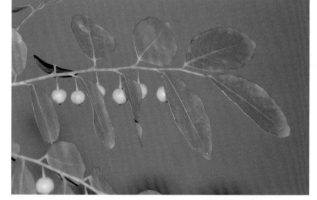

果枝　Fruiting branch
摄影：杨庆松　Photo by: Yang Qingsong

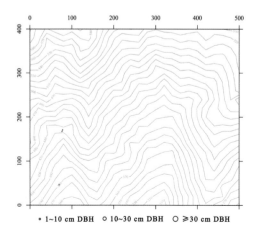

个体分布图 Distribution of individuals

径级分布表 DBH class

胸径区间 (Diameter class) (cm)	个体数 (No. of individuals in the plot)	比例 (Proportion) (%)
1～2	1	33.33
2～5	1	33.33
5～10	1	33.33
10～20	0	0.00
20～30	0	0.00
30～60	0	0.00
≥60	0	0.00

72 油桐　　　　　　　　　　　　　　　　　　　　Yóu tóng | Tung-Oil Tree

Vernicia fordii (Hemsl.) Airy Shaw
大戟科 | Euphorbiaceae

代码 (SpCode) = VERFOR
个体数 (Individual number/20 hm^2) = 309
最大胸径 (Max DBH) = 29.6 cm
重要值排序 (Importance value rank) = 31

落叶乔木，高达10 m；树皮灰色，近光滑；枝条轮生，粗壮，无毛，具明显皮孔。叶卵圆形，顶端短尖，基部截平至浅心形，全缘，稀1~3浅裂；叶柄与叶片近等长，几无毛，顶端有2枚扁平、无柄腺体。花雌雄同株，先叶或与叶同时开放。核果近球状，直径4~6(~8) cm，种子3~4(~8) 颗。花期3~4月，果期8~11月。

Deciduous trees, up to 10 m tall, monoecious; bark gray, nearly smooth. Branches verticillate, stout, glabrous, prominently lenticellate. Leaf blade ovate, base truncate to shallowly cordate, margin entire, rarely shallowly 1~3-fid, apex acute, petiole as long as leaf blade, glabrescent, with 2 compressed and sessile glands. Inflorescences flat-topped panicles of cymes, appearing generally before new leaves, usually bisexual. Drupe subglobose, 4~6(~8) cm in diam., 3- or 4(~8) -seeded. Fl. Mar. - Apr., fr. Aug. - Nov..

树干　　Trunk
摄影：杨庆松　Photo by: Yang Qingsong

枝叶　　Branch and leaves
摄影：杨庆松　Photo by: Yang Qingsong

花枝　　Flowering branch
摄影：杨庆松　Photo by: Yang Qingsong

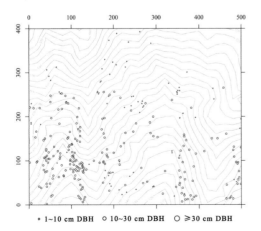

个体分布图　Distribution of individuals

径级分布表 DBH class

胸径等级 (Diameter class) (cm)	个体数 (No. of individuals in the plot)	比例 (Proportion) (%)
1~2	39	12.62
2~5	17	5.50
5~10	32	10.36
10~20	165	53.40
20~30	56	18.12
30~60	0	0.00
≥60	0	0.00

73 虎皮楠

Hǔ pí nán | Tigernanmu

Daphniphyllum oldhami (Hemsl.) Rosenth.
交让木科 | Daphniphyllaceae

代码 (SpCode) = DAPOLD
个体数 (Individual number/20 hm^2) = 462
最大胸径 (Max DBH) = 59.0 cm
重要值排序 (Importance value rank) = 29

常绿乔木，高达15 m。小枝纤细，暗褐色。叶纸质，披针形或倒卵状披针形或长圆形或长圆状披针形，长9～14 cm，宽2.5～4 cm，先端急尖或渐尖或短尾尖，基部楔形或钝，边缘反卷，干后叶面暗绿色，具光泽，叶背通常显著被白粉；叶柄长2～3.5 cm，纤细，上面具槽。果椭圆或倒卵圆形，暗褐至黑色。花期3～5月，果期8～11月。

Evergreen trees, up to 15 m tall; branchlets slender, dark brown. Petiole 2~3.5 cm, slender; leaf blade lanceolate, obovate-lanceolate, oblong, or oblong-lanceolate, 9~14 × 2.5~4 cm, adaxially dark green in dried state, base cuneate or obtuse, margins revolute, apex acute, acuminate, or shortly caudate. Drupe ellipsoidal or obovate-globose, dark brown. Fl. Mar. - May, fr. Aug. - Nov..

幼苗　Seedling
摄影：刘何铭　Photo by: Liu Heming

花枝　Flowering branches
摄影：杨庆松　Photo by: Yang Qingsong

果枝　Fruiting branches
摄影：杨庆松　Photo by: Yang Qingsong

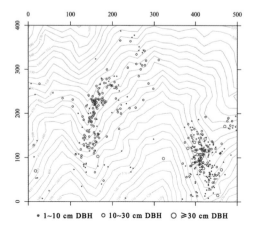

个体分布图 Distribution of individuals

径级分布表 DBH class

胸径区间 (Diameter class) (cm)	个体数 (No. of individuals in the plot)	比例 (Proportion) (%)
1～2	37	8.01
2～5	78	16.88
5～10	136	29.44
10～20	157	33.98
20～30	35	7.58
30～60	19	4.11
≥60	0	0.00

74 南酸枣

Nán suān zǎo | Axillary Southern Wildjujube

Choerospondias axillaris (Roxb.) Burtt et Hill
漆树科 | Anacardiaceae

代码 (SpCode) = CHOAXI
个体数 (Individual number/20 hm^2) = 1337
最大胸径 (Max DBH) = 87.5 cm
重要值排序 (Importance value rank) = 3

落叶乔木，8~20 m高。小枝暗紫褐色，无毛，具皮孔。叶柄基部膨大；奇数羽状复叶，具小叶3~6对；小叶片卵形到卵状披针形或长圆状卵形，纸质，无毛或背面脉腋有簇毛在，基部偏斜。雌雄同株，雌花比雄花大，单生在上部叶的叶腋。核果椭圆形或倒卵状椭圆形，成熟时黄色。花期4~5月，果期8~11月。

Deciduous trees, 8~20 m tall. Branchlets dark purplish brown, glabrous, lenticellate. Petiole inflated at base; leaf imparipinnately compound, with 3~6 leaflets; leaflet blade ovate to ovate-lanceolate or oblong-ovate, papery, glabrous or abaxially with tufts of hair in vein axils, base oblique. Androgynous, female flowers solitary in axils of distal leaves, larger than male flowers; ovary ca. 1.5 mm, 5-locular, style ca. 0.5 mm. Drupe ellipsoidal or obovate-ellipsoidal, yellow at maturity. Fl. Apr. - May, fr. Aug. - Nov..

幼苗　　Seedling
摄影：刘何铭　Photo by: Liu Heming

花枝　　Flowering branch
摄影：杨庆松　Photo by: Yang Qingsong

果枝　　Fruiting branch
摄影：杨庆松　Photo by: Yang Qingsong

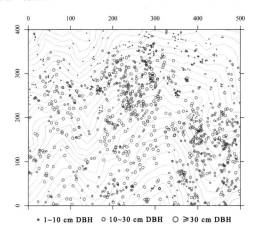

个体分布图 Distribution of individuals

径级分布表 DBH class

胸径等级 (Diameter class) (cm)	个体数 (No. of individuals in the plot)	比例 (Proportion) (%)
1~2	250	18.70
2~5	154	11.52
5~10	37	2.77
10~20	161	12.04
20~30	335	25.06
30~60	377	28.20
≥60	23	1.72

75 盐肤木

Yán fū mù | China Sumac

Rhus chinensis Mill.
漆树科 | Anacardiaceae

代码 (SpCode) = RHUCHI
个体数 (Individual number/20 hm^2) = 2
最大胸径 (Max DBH) = 1.3 cm
重要值排序 (Importance value rank) = 144

落叶灌木到乔木，高2~10 m；小枝被锈色短柔毛，具皮孔。叶无柄，奇数羽状复叶；叶轴具宽的叶状翅，被锈色短柔毛；小叶 (5) 7~13；小叶叶片卵形到长圆形，靠近先端小叶更大，正面深绿色，疏生短柔毛或近无毛，背面淡绿色，有白粉及锈色短柔毛。花序多分枝，密被锈色短柔毛。核果球状，成熟时红色。花期8-9月，果期10月。

Deciduous shrubs to trees, 2~10 m tall; branchlets ferruginous pubescent, lenticellate. Leaf blade sessile, imparipinnately compound; rachis broadly winged to wingless, ferruginous pubescent; leaflets (5~) 7~13; leaflet blade ovate to oblong, increasing in size toward apex, 6~12 × 3~7 cm, adaxially dark green, sparsely pubescent or glabrescent, abaxially lighter green, glaucous, and ferruginous pubescent. Inflorescence many branched, densely ferruginous pubescent. Drupe globose, red at maturity. Fl. Aug. - Sep., fr. Oct..

复叶   Compound leaf
摄影：杨庆松   Photo by: Yang Qingsong

花枝   Flowering branch
摄影：杨庆松   Photo by: Yang Qingsong

果   Fruits
摄影：杨庆松   Photo by: Yang Qingsong

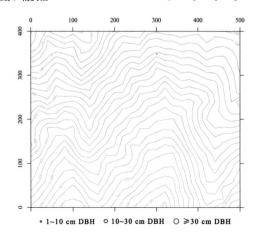

个体分布图 Distribution of individuals

径级分布表 DBH class

胸径区间 (Diameter class) (cm)	个体数 (No. of individuals in the plot)	比例 (Proportion) (%)
1~2	2	100.00
2~5	0	0.00
5~10	0	0.00
10~20	0	0.00
20~30	0	0.00
30~60	0	0.00
≥60	0	0.00

76 野漆　　　　　　　　　　　　　　　　　　　　　　Yě qī | Field Lacquertree

Toxicodendron succedaneum (Linn.) O. Kuntze
漆树科 | Anacardiaceae

代码 (SpCode) = TOXSUC
个体数 (Individual number/20 hm^2) = 103
最大胸径 (Max DBH) = 29.9 cm
重要值排序 (Importance value rank) = 68

落叶乔木，高可达10 m；小枝粗壮无毛。叶柄6～9 cm；叶片奇数羽状复叶，20～35 cm；小叶5～15，对生或近对生；小叶叶柄不明显或2～5 mm；小叶叶片纸质或薄革质，两面无毛到疏生短柔毛，背面有白霜，基部偏斜，圆形或宽楔形，边缘全缘。核果偏斜，直径7～10 mm，压扁。花期3～5月，果期8～10月。

Deciduous trees, up to 10 m tall; branchlets glabrous and strong. Petiole 6~9 cm; leaf blade imparipinnately compound, 20~35 cm; leaflets 5~15, opposite or subopposite; leaflet petiolule indistinct or 2~5 mm; leaflet papery or thinly leathery, glabrous to sparsely pubescent on both surfaces, glaucous abaxially, base oblique, rounded or broadly cuneate, margin entire. Drupe, asymmetrical, 7~10 mm in diam., compressed. Fl. Mar. - May, fr. Aug. - Oct..

叶背　Leaf abaxial surface
摄影：杨庆松　Photo by: Yang Qingsong

枝叶　Branch and leaves
摄影：杨庆松　Photo by: Yang Qingsong

果枝　Fruiting branches
摄影：杨庆松　Photo by: Yang Qingsong

径级分布表 DBH class

胸径等级 (Diameter class) (cm)	个体数 (No. of individuals in the plot)	比例 (Proportion) (%)
1～2	24	23.30
2～5	24	23.30
5～10	35	33.98
10～20	16	15.53
20～30	4	3.88
30～60	0	0.00
≥60	0	0.00

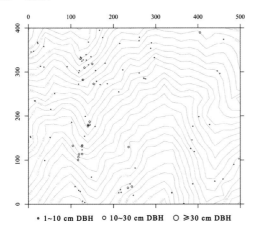
个体分布图　Distribution of individuals

77 木蜡树

Mù là shù | Hairyfruit Lacquertree

Toxicodendron sylvestre (Sieb. et Zucc.) O. Kuntze
漆树科 | Anacardiaceae

代码 (SpCode) = TOXSYL
个体数 (Individual number/20 hm^2) = 7
最大胸径 (Max DBH) = 7.0 cm
重要值排序 (Importance value rank) = 120

落叶乔木或小乔木，高10 m；幼枝和芽被黄棕色绒毛。叶柄4～8 cm，叶柄和叶轴密被黄色绒毛；叶片奇数羽状复叶，18～30 cm；小叶7～15，对生；小叶纸质，背面密被短柔毛，基部偏斜，圆形或宽楔形，边缘全缘，先端渐尖或锐尖。核果极偏斜，压扁。花期4～5月，果期6～10月。

Deciduous trees or small trees, to 10 m tall; young branchlets and terminal buds yellowish brown tomentose. Petiole 4~8 cm, petiole and rachis densely yellow tomentose; leaf blade imparipinnately compound, 18~30 cm; leaflets 7~15, opposite; leaflet papery, abaxially densely pubescent, base oblique, rounded or broadly cuneate, margin entire, apex acuminate or acute. Drupe oblique, compressed. Fl. Apr. - May, fr. Jun. - Oct..

枝叶 Branches and leaves
摄影：杨庆松 Photo by: Yang Qingsong

叶 Leaves
摄影：杨庆松 Photo by: Yang Qingsong

果枝 Fruiting branches
摄影：杨庆松 Photo by: Yang Qingsong

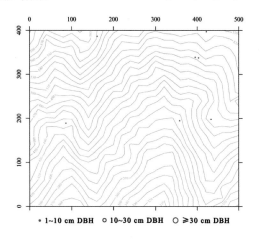

个体分布图 Distribution of individuals

径级分布表 DBH class

胸径区间 (Diameter class) (cm)	个体数 (No. of individuals in the plot)	比例 (Proportion) (%)
1～2	3	42.86
2～5	3	42.86
5～10	1	14.29
10～20	0	0.00
20～30	0	0.00
30～60	0	0.00
≥60	0	0.00

78 短梗冬青

Duǎn gěng dōng qīng | Buerger Holly

Ilex buergeri Miq.
冬青科 | Aquifoliaceae

代码 (SpCode) = ILEBUE
个体数 (Individual number/20 hm²) = 716
最大胸径 (Max DBH) = 54.1 cm
重要值排序 (Importance value rank) = 27

常绿乔木，高7~15 m；树皮黑色，光滑。小枝圆柱状，纵向脊和具槽，密被短柔毛；老枝无毛。叶柄4~8 mm，上面具槽，被短柔毛；叶片正面深绿色，有光泽，卵形、长圆形或卵状披针形，革质，基部圆形、钝或楔形，边缘稍反卷，疏生不规则浅锯齿，先端渐尖。果红色，球形或近球形。花期4~6月，果期10~11月。

Evergreen trees, 7~15 m tall; bark black, smooth. Branchlets terete, longitudinally ridged and sulcate, densely pubescent; old branchlets glabrescent. Petiole 4~8 mm, adaxially sulcate, pubescent; leaf blade adaxially dark green, shiny, ovate, oblong, or ovate-lanceolate, leathery, base rounded, obtuse, or cuneate, margin slightly reflexed, sparsely irregularly shallowly serrate, apex acuminate. Fruit red, globose or subglobose. Fl. Apr. - Jun., fr. Oct. - Nov..

树干　　Trunk
摄影：杨庆松　Photo by: Yang Qingsong

叶　　Leaves
摄影：杨庆松　Photo by: Yang Qingsong

果枝　　Fruiting branch
摄影：杨庆松　Photo by: Yang Qingsong

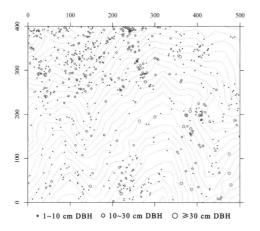
个体分布图　Distribution of individuals

径级分布表 DBH class

胸径等级 (Diameter class) (cm)	个体数 (No. of individuals in the plot)	比例 (Proportion) (%)
1~2	232	32.40
2~5	174	24.30
5~10	108	15.08
10~20	141	19.69
20~30	42	5.87
30~60	19	2.65
≥60	0	0.00

79 冬青 Dōng qīng | Chinese Holly

Ilex chinensis Sims
冬青科 | Aquifoliaceae

代码 (SpCode) = ILECHI

个体数 (Individual number/20 hm^2) = 12

最大胸径 (Max DBH) = 29.4 cm

重要值排序 (Importance value rank) = 97

常绿乔木，高15 m。植株无毛，有时在雄株的芽、叶柄和幼叶的正面中脉具长柔毛。叶柄8～10 mm，扁平或狭具槽；叶片深褐色，薄革质至革质，无毛，基部楔形或钝，边缘具圆齿，或幼叶有锯齿，先端渐尖。花序聚伞花序，单生，在当年小枝上腋生，花淡紫色或紫红色，无毛。果实红色，狭球状。花期4～7月，果期7～12月。

Evergreen trees, to 15 m tall. Plants glabrous, or sometimes villous on terminal buds, petioles, and adaxial midvein of young leaves of male plant. Petiole 8~10 mm, flat or narrowly sulcate adaxially; leaf blade deep brown, thinly leathery to leathery, glabrous, base cuneate or obtuse, margin crenate, or sometimes young leaf serrate, apex acuminate. Inflorescences: cymes, solitary, axillary on current year's branchlets, flowers purplish or purple-red, glabrous. Fruit red, narrowly globose. Fl. Apr. - Jul., fr. Jul. - Dec..

枝叶　Branchs and leaves
摄影：杨庆松　Photo by: Yang Qingsong

花枝　Flowering branch
摄影：杨庆松　Photo by: Yang Qingsong

果枝　Fruiting branches
摄影：徐明杰　Photo by: Xu Mingjie

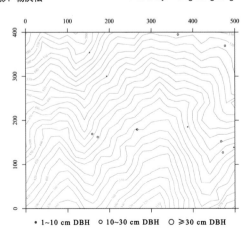

个体分布图　Distribution of individuals

径级分布表　DBH class

胸径区间 (Diameter class) (cm)	个体数 (No. of individuals in the plot)	比例 (Proportion) (%)
1~2	4	33.33
2~5	1	8.33
5~10	0	0.00
10~20	4	33.33
20~30	3	25.00
30~60	0	0.00
≥60	0	0.00

80 细刺枸骨 Xì cì gǒu gǔ | Slenderprickle Holly

Ilex hylonoma Hu et Tang
冬青科 | Aquifoliaceae

代码 (SpCode) = ILEHYL
个体数 (Individual number/20 hm^2) = 13
最大胸径 (Max DBH) = 10.1 cm
重要值排序 (Importance value rank) = 109

常绿乔木，高达10 m。小枝栗色，直，无毛或变无毛。叶柄8~14 mm，背面具皱纹，正面具槽，稍具短柔毛或无毛；叶片薄到厚革质，中脉正面凹陷，具柔毛，后脱落无毛，基部钝或锐，很少楔形，边缘粗而尖的锯齿，有时齿尖为弱刺。花淡黄色，果成熟时棕红色，椭圆形近球形。花期3~5月，果期7~11月。

Evergreen trees, to 10 m tall. Branchlets castaneous, straight, glabrescent or glabrous. Petiole 8~14 mm, abaxially rugose, adaxially sulcate and minutely pubescent or glabrous; leaf blade thinly to thickly leathery, midvein adaxially impressed, pilose, glabrescent, or glabrous, base obtuse or acute, rarely cuneate, margin coarsely and sharply serrate, sometimes teeth ending in weak spines. Flowers 4-merous, yellowish Fruit brown, ellipsoid-subglobose. Fl. Mar. - May, fr. Jul. - Nov..

树干　Trunk
摄影：杨庆松　Photo by: Yang Qingsong

枝叶　Branch and leaves
摄影：杨庆松　Photo by: Yang Qingsong

叶背　Leaf abaxial surface
摄影：杨庆松　Photo by: Yang Qingsong

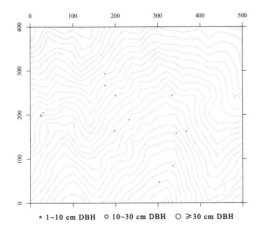

个体分布图　Distribution of individuals

径级分布表 DBH class

胸径等级 (Diameter class) (cm)	个体数 (No. of individuals in the plot)	比例 (Proportion) (%)
1~2	2	15.38
2~5	2	15.38
5~10	8	61.54
10~20	1	7.69
20~30	0	0.00
30~60	0	0.00
≥60	0	0.00

81 皱柄冬青

Zhòu bǐng dōng qīng | Keng Holly

Ilex kengii S. Y. Hu
冬青科 | Aquifoliaceae

代码 (SpCode) = ILEKEN
个体数 (Individual number/20 hm^2) = 72
最大胸径 (Max DBH) = 23.9 cm
重要值排序 (Importance value rank) = 76

常绿乔木，4~13 (15) m；树皮灰。小枝褐色，纤细，无毛或疏生微柔毛；当年小枝具纵棱和具槽。叶柄7~15 (18) mm，横向具皱纹；叶片薄革质或革质，叶背面多具棕色腺点，两面无毛，基部钝、圆形或宽楔形，全缘，有时稍外卷。果红色，球状，直径3~5 mm。花期5月，果期6~11月。

Evergreen trees, 4~13(15) m tall; bark gray. Branchlets brown, slender, glabrous or sparsely puberulent; current year's branchlets longitudinally angular and sulcate. Petiole 7~15(18) mm, abaxially transversely rugose; leaf blade thinly leathery or leathery, abaxially brown glandular punctate, rarely not punctate, both surfaces glabrous, base obtuse, rounded, or broadly cuneate, margin entire, sometimes slightly revolute. Fruit red, globose, 3~5 mm in diam. Fl. May, fr. Jun. - Nov..

树干　Trunk
摄影：杨庆松　Photo by: Yang Qingsong

枝叶　Branches and leaves
摄影：杨庆松　Photo by: Yang Qingsong

果枝　Fruiting branches
摄影：徐明杰　Photo by: Xu Mingjie

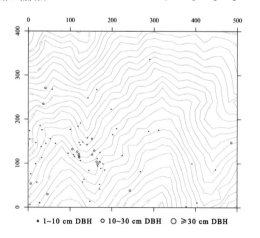

个体分布图　Distribution of individuals

径级分布表 DBH class

胸径区间 (Diameter class) (cm)	个体数 (No. of individuals in the plot)	比例 (Proportion) (%)
1~2	26	36.11
2~5	12	16.67
5~10	17	23.61
10~20	14	19.44
20~30	3	4.17
30~60	0	0.00
≥60	0	0.00

82 大叶冬青

Dà yè dōng qīng | Broadleaf Holly

Ilex latifolia Thunb.
冬青科 | Aquifoliaceae

代码 (SpCode) = ILELAT
个体数 (Individual number/20 hm^2) = 111
最大胸径 (Max DBH) = 47.7 cm
重要值排序 (Importance value rank) = 48

常绿乔木，高20 m，全株无毛；树皮灰黑色，光滑。枝黄棕色或棕色，粗壮，纵向脊和具槽，光滑。叶柄近圆柱状，1.5～2.5 cm；叶厚革质，中脉上面凹陷，背面侧脉不明显，正面明显，基部圆形或宽楔形，边缘疏生锯齿，齿黑尖，先端钝或短渐尖。果红色，球状，直径约7 mm。花期4～5月，果期9～10月。

Evergreen trees, to 20 m tall, glabrous throughout; bark gray-black, smooth. Branches yellow-brown or brown, strong, longitudinally ridged and sulcate, smooth; petiole subterete, 1.5~2.5 cm; leaf blade thickly leathery, midvein impressed adaxially, lateral veins obscure abaxially, obvious adaxially, base rounded or broadly cuneate, margin sparsely serrate, teeth black at tips, apex obtuse or shortly acuminate. Fruit red, globose, ca. 7 mm in diam. Fl. Apr. - May, fr. Sep. - Oct..

叶 Leaf
摄影：杨庆松 Photo by: Yang Qingsong

花枝 Flowering branch
摄影：王樟华 Photo by: Wang Zhanghua

果枝 Fruiting branches
摄影：王樟华 Photo by: Wang Zhanghua

径级分布表 DBH class

胸径等级 (Diameter class) (cm)	个体数 (No. of individuals in the plot)	比例 (Proportion) (%)
1～2	9	8.11
2～5	26	23.42
5～10	32	28.83
10～20	22	19.82
20～30	13	11.71
30～60	9	8.11
≥60	0	0.00

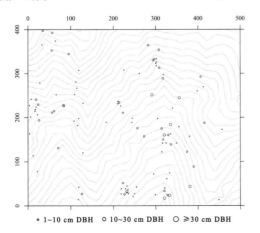
个体分布图 Distribution of individuals

83 小果冬青

Xiǎo guǒ dōng qīng | Smallfruit Holly

Ilex micrococca Maxim.
冬青科 | Aquifoliaceae

代码 (SpCode) = ILEMIC
个体数 (Individual number/20 hm^2) = 22
最大胸径 (Max DBH) = 49.0 cm
重要值排序 (Importance value rank) = 82

落叶乔木，高12～20 m。小枝粗壮，无毛，具白色、圆形或长圆形常并生的气孔。叶柄纤细，无毛；叶面深绿色，背面淡绿色，卵形、卵状椭圆形，或卵形长圆形，基部圆形或宽楔形，常偏斜，边缘近全缘或具芒状锯齿，先端长渐尖。果红色或黄色，球状，直径约3 mm。花期5～6月，果期9～11月。

Deciduous trees, 12~20 m tall. Branchlets thick, glabrous or pubescent, with conspicuous, large, circular or oblong, often coalescent white lenticels. Petiole slender, glabrous; leaf blade abaxially greenish, adaxially deep green, ovate, ovate-elliptic, or ovate-oblong, base rounded or broadly cuneate, often oblique, margin subentire or aristate- serrate, apex long acuminate. Fruit red or yellow, globose, ca. 3 mm in diam. Fl. May - Jun., fr. Sep. - Nov..

树干　Trunk
摄影：杨庆松　Photo by: Yang Qingsong

果枝　Fruiting branches
摄影：杨庆松　Photo by: Yang Qingsong

果　Fruits
摄影：杨庆松　Photo by: Yang Qingsong

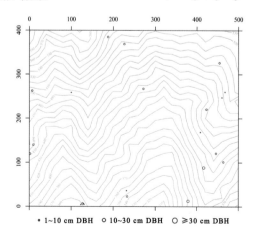

个体分布图　Distribution of individuals

径级分布表 DBH class

胸径区间 (Diameter class) (cm)	个体数 (No. of individuals in the plot)	比例 (Proportion) (%)
1～2	0	0.00
2～5	2	9.09
5～10	6	27.27
10～20	8	36.36
20～30	4	18.18
30～60	2	9.09
≥60	0	0.00

84 铁冬青

Tiě dōng qīng | Iron Holly

Ilex rotunda Thunb.
冬青科 | Aquifoliaceae

代码 (SpCode) = ILEROT
个体数 (Individual number/20 hm^2) = 165
最大胸径 (Max DBH) = 36.4 cm
重要值排序 (Importance value rank) = 49

常绿乔木，高达20 m；树皮灰色至灰黑色。幼枝紫褐色，具纵棱，无毛，很少被微柔毛；叶柄狭具槽；叶片卵形、倒卵形或椭圆形，薄革质或纸质，两面无毛，中脉在上面凹陷，下面中脉靠近叶柄处常带紫褐色，基部钝或楔形，边缘全缘，稍下弯，先端短渐尖。果红色，近球形，直径的4～6 mm。花期4～6月，果期8～12月。

Evergreen trees, to 20 m tall; bark gray to gray-black. Young branchlets purple brown, longitudinally angular, glabrous, rarely puberulent. Petiole narrowly sulcate adaxially; leaf blade ovate, obovate, or elliptic, thinly leathery or papery, both surfaces glabrous, midvein impressed adaxially, midvein abaxially usually purple brown near to petiole, base obtuse or cuneate, margin entire, slightly recurved, apex shortly acuminate. Fruit red, subglobose, rarely ellipsoid, 4~6 mm in diam. Fl. Apr. - Jun., fr. Aug. - Dec..

枝叶　Branch and leaves
摄影：杨庆松　Photo by: Yang Qingsong

雌花序　Female inflorescences
摄影：汪远　Photo by: Wang yuan

果　Fruits
摄影：杨庆松　Photo by: Yang Qingsong

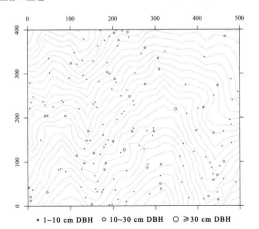

个体分布图　Distribution of individuals

径级分布表　DBH class

胸径等级 (Diameter class) (cm)	个体数 (No. of individuals in the plot)	比例 (Proportion) (%)
1～2	42	25.45
2～5	33	20.00
5～10	47	28.48
10～20	30	18.18
20～30	11	6.67
30～60	2	1.21
≥60	0	0.00

85 百齿卫矛

Bǎi chǐ wèi máo | Hundredtooth Euonymus

Euonymus centidens H. Lév.
卫矛科 | Celastraceae

代码 (SpCode) = EUOCEN
个体数 (Individual number/20 hm^2) = 4
最大胸径 (Max DBH) = 1.1 cm
重要值排序 (Importance value rank) = 133

常绿灌木，2~4 (5) m；小枝四棱形。叶柄近无或有短柄，少于约3 mm；叶片薄革质或厚纸质，倒卵形或椭圆状倒卵形，6~11×2.5~4.5 cm，基部楔形或渐狭，边缘有细锯齿，先端锐尖、渐尖或尾状。蒴果4深裂，种子长圆状，假种皮黄红色。花期5~6月，果9~11月。

Evergreen shrubs, 2~4 (~5) m tall; branchlets 4-angled. Petiole sessile or very short, less than ca. 3 mm; leaf blade thinly leathery or thickly papery, obovate or elliptic-obovate, 6~11 × 2.5~4.5 cm, base cuneate or attenuate, serrulate to serrate, apex acute, acuminate, or caudate. Capsule reddish brown when fresh, dark brown or gray when dry, 4-lobed, Aril bright red. Fl. May - Jun., fr. Sep. - Nov..

幼苗　Seedlings
摄影：杨庆松　Photo by: Yang Qingsong

枝叶　Branch and leaves
摄影：杨庆松　Photo by: Yang Qingsong

果枝　Fruiting branches
摄影：杨庆松　Photo by: Yang Qingsong

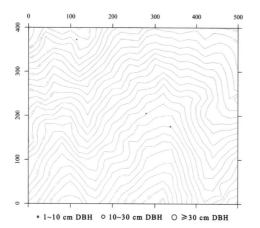
个体分布图 Distribution of individuals

径级分布表 DBH class

胸径区间 (Diameter class) (cm)	个体数 (No. of individuals in the plot)	比例 (Proportion) (%)
1~2	4	100.00
2~5	0	0.00
5~10	0	0.00
10~20	0	0.00
20~30	0	0.00
30~60	0	0.00
≥60	0	0.00

86 中华卫矛

Zhōng huá wèi máo | China Euonymus

Euonymus nitidus Benth.
卫矛科 | Celastraceae

代码 (SpCode) = EUONIT
个体数 (Individual number/20 hm^2) = 15
最大胸径 (Max DBH) = 9.8 cm
重要值排序 (Importance value rank) = 112

常绿灌木至小乔木，2~10 m高。枝灰黑灰棕色，圆柱状，粗壮，小枝绿色或黄绿色，具条纹。叶柄5~8 (12) mm；叶片椭圆形或长圆状椭圆形，6.5~10 (15) × 3~4 (6) cm，基部楔形或渐尖，边缘全缘具细圆齿，先端锐尖或渐尖，有时短尾状。蒴果四棱形，假种皮黄橙色，全包种子，上部两侧开裂。花期5~7月，果期7~翌年1月。

Evergreen shrubs to small trees, 2~10 m tall; branches gray-black to gray-brown, terete, sturdy, twigs greenish or yellow-greenish, striate. Petiole 5~8(~12) mm; leaf blade elliptic or oblong-elliptic, 6.5~10(~15) × 3~4(~6) cm, base cuneate or acuminate, margin entire to crenulate, apex acute or acuminate, sometimes shortly caudate. Capsule 4-angled, aril orange . Fl. Mar. - Jul., fr. Jul. - Jan. of following year.

小枝 Branches
摄影：杨庆松 Photo by: Yang Qingsong

叶和果 Leaves and Fruits
摄影：杨庆松、杜诚 Photo by: Yang Qingsong, Du Cheng

花枝 Flowering branches
摄影：杜诚 Photo by: Du Cheng

径级分布表 DBH class

胸径等级 (Diameter class) (cm)	个体数 (No. of individuals in the plot)	比例 (Proportion) (%)
1~2	5	33.33
2~5	8	53.33
5~10	2	13.33
10~20	0	0.00
20~30	0	0.00
30~60	0	0.00
≥60	0	0.00

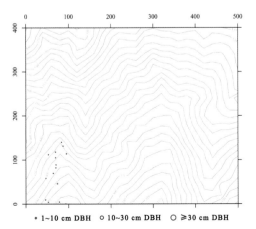

个体分布图 Distribution of individuals

87 野鸦椿 Yě yā chūn | Common Euscaphis

Euscaphis japonica (Thunb.) Kanitz
省沽油科 | Staphyleaceae

代码 (SpCode) = EUSJAP
个体数 (Individual number/20 hm^2) = 18
最大胸径 (Max DBH) = 12.3 cm
重要值排序 (Importance value rank) = 105

落叶小乔木或灌木，可达6～8 m高；树皮灰褐色，有纵条纹。小枝和芽红紫色。奇数羽状复叶，对生，叶揉碎后有难闻的气味；小叶柄1～2 mm，脱落；小叶叶片椭圆形至长圆状卵形或长圆状披针形，很少卵形，纸质，无毛或具柔毛沿脉，边缘疏生细锯齿具牙齿，先端渐尖。蓇葖果，果皮紫红色，种子近球形，直径约5 mm；假种皮肉质，亮黑色。花期4～6月，果期8～11月。

Small deciduous trees or shrubs, 6~8 m tall; bark grayish brown, striped. Twigs and buds dark purple. Leaf blade imparipinnately compound, opposite, with unpleasant odor when crushed; petiolule 1~2 mm; leaflet blades elliptic to oblong-ovate or sometimes oblong-lanceolate, rarely ovate, papery, glabrous or pilose along veins, margin sparsely serrulate with glandular teeth, apex acuminate. Seeds shining black, subglobose, ca. 5 mm in diam. Fl. Apr. - Jun., fr. Aug.-Nov..

枝叶　Branch and leaves
摄影：杨庆松　Photo by: Yang Qingsong

果枝　Fruiting branches
摄影：杨庆松　Photo by: Yang Qingsong

果　Fruits
摄影：胡瑾瑾　Photo by: Hu Jinjin

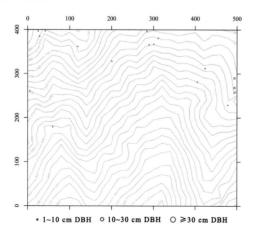

个体分布图 Distribution of individuals

径级分布表 DBH class

胸径区间 (Diameter class) (cm)	个体数 (No. of individuals in the plot)	比例 (Proportion) (%)
1～2	9	50.00
2～5	3	16.67
5～10	4	22.22
10～20	2	11.11
20～30	0	0.00
30～60	0	0.00
≥60	0	0.00

88 锐角槭 | Ruì jiǎo qì | Acute Maple

Acer acutum W. P. Fang
槭树科 | Aceraceae

代码 (SpCode) = ACEACU
个体数 (Individual number/20 hm^2) = 47
最大胸径 (Max DBH) = 45.5 cm
重要值排序 (Importance value rank) = 59

落叶乔木，高10 m，雌雄同株，树皮褐色或棕灰色。小枝平滑，无毛。叶柄4～12 cm，幼时近先端被微柔毛，后脱落；叶片纸质，正面无毛，背面幼时具短柔毛，基部心形或近心形，(5或) 7浅裂；裂片宽卵形或三角形，中部裂片与侧裂片通常渐尖。果淡褐色，无毛；翅长圆形。花期4月，果期8月。

Deciduous trees to 10 m tall, andromonoecious. Bark brown or brownish gray. Branchlets smooth, glabrous. Petiole 4~12 cm, puberulent near apex when young, glabrescent; leaf blade papery, abaxially pubescent when young, adaxially glabrous, base cordate or subcordate, (5 or) 7-lobed; lobes broadly ovate or triangular, middle lobe and lateral lobes usually acuminate apically. Fruit pale brown, glabrous; samara wing oblong. Fl. Apr., fr. Aug..

树干　Trunk
摄影：杨庆松　Photo by: Yang Qingsong

枝叶　Branch and leaves
摄影：杨庆松　Photo by: Yang Qingsong

老叶　Old leaf
摄影：杨庆松　Photo by: Yang Qingsong

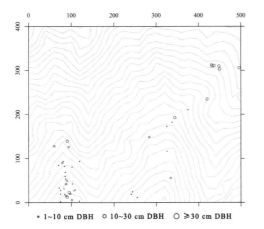
个体分布图　Distribution of individuals

径级分布表　DBH class

胸径等级 (Diameter class) (cm)	个体数 (No. of individuals in the plot)	比例 (Proportion) (%)
1～2	5	10.64
2～5	15	31.91
5～10	6	12.77
10～20	6	12.77
20～30	4	8.51
30～60	11	23.40
≥60	0	0.00

89 三角槭

Sān jiǎo qì | Buerger Maple

Acer buergerianum Miq.
槭树科 | Aceraceae

代码 (SpCode) = ACEBUE
个体数 (Individual number/20 hm^2) = 5
最大胸径 (Max DBH) = 66.7 cm
重要值排序 (Importance value rank) = 85

落叶乔木，5～20 m高，雌雄同株。树皮粗糙，小枝纤细。叶柄2.5～5 (8) cm，无毛；叶片具3主脉，基部圆形或楔形，浅或3深裂；中裂片三角状卵形或披针形，先端锐尖或短渐尖；侧裂片短，边缘通常全缘，偶有疏齿，先端锐尖。果黄褐色；翅中部最宽，基部狭窄。花期4月，果期8月。

Deciduous trees 5~20 m tall, andromonoecious. Bark rough, branchlets slender. Petiole 2.5~5(~8) cm, glabrous; leaf blade with 3 primary veins, base rounded or cuneate, shallowly or deeply 3-lobed; middle lobe triangular-ovate or lanceolate, apex acute or shortly acuminate; lateral lobes short, margin usually entire, rarely with a few serrations, apex acute, sometimes lobes very small or obsolete. Fruit yellowish brown; wing broad at middle, contracted at base. Fl. Apr., fr. Aug..

果枝　Fruiting branches
摄影：杨庆松　Photo by: Yang Qingsong

叶　Leaves
摄影：杨庆松　Photo by: Yang Qingsong

花枝　Flowering branches
摄影：杨庆松　Photo by: Yang Qingsong

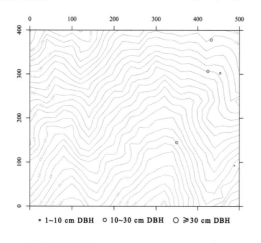

个体分布图　Distribution of individuals

径级分布表　DBH class

胸径区间 (Diameter class) (cm)	个体数 (No. of individuals in the plot)	比例 (Proportion) (%)
1～2	0	0.00
2～5	1	20.00
5～10	0	0.00
10～20	0	0.00
20～30	1	20.00
30～60	2	40.00
≥60	1	20.00

90 毛脉槭

Máo mài qì | Hairyvein Maple

Acer pubinerve Rehd.
槭树科 | Aceraceae

代码 (SpCode) = ACEPUB
个体数 (Individual number/20 hm^2) = 475
最大胸径 (Max DBH) = 50.1 cm
重要值排序 (Importance value rank) = 19

落叶乔木，高可达15 m。树皮深灰色，光滑，小枝绿色。叶柄4~5 cm，密被短绒毛；叶片背面近无毛，在脉上密被长柔毛，正面除脉上被短柔毛外无毛，5浅裂，边缘稍有短尖锯齿，近基部通常全缘，近心形，先端尾状。果黄色，翅张开成钝角或近水平。花期4月，果期10月。

Deciduous trees, up to 15 m tall. Bark dark gray, smooth. Branchlets green, glabrous or densely pubescent. Petiole 4~5 cm, densely velutinous; leaf blade abaxially nearly glabrous except densely villous on veins, adaxially glabrous except shortly velutinous on veins, 5-lobed, margin slightly serrate, with short acuminate teeth, usually entire toward base, base subcordate, apex caudate. Fruit yellowish, wings spreading obtusely or nearly horizontally. Fl. Apr., fr. Oct..

幼苗　Seedling
摄影：刘何铭　Photo by: Liu Heming

花枝　Flowering branches
摄影：杨庆松　Photo by: Yang Qingsong

果　Fruits
摄影：杨庆松　Photo by: Yang Qingsong

径级分布表 DBH class

胸径等级 (Diameter class) (cm)	个体数 (No. of individuals in the plot)	比例 (Proportion) (%)
1~2	135	28.42
2~5	56	11.79
5~10	34	7.16
10~20	79	16.63
20~30	103	21.68
30~60	68	14.32
≥60	0	0.00

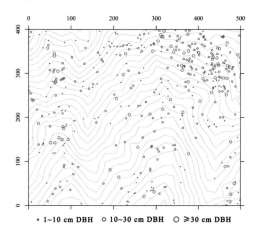

个体分布图 Distribution of individuals

91 无患子

Wú huàn zǐ | China Soapberry

Sapindus saponaria Linn.
无患子科 | Sapindaceae

代码 (SpCode) = SAPSAP
个体数 (Individual number/20 hm^2) = 149
最大胸径 (Max DBH) = 50.0 cm
重要值排序 (Importance value rank) = 39

落叶乔木，高达20 m；树皮淡灰棕色或黑褐色。幼枝绿色，无毛。叶柄25～45 cm或更长，轴稍平，正面具槽，无毛或具小柔毛；小叶5～8对，通常近对生；小叶柄长约5 mm；叶片正面发亮，狭椭圆状披针形或稍镰刀形，薄纸质，基部楔形，稍不对称，先端锐尖或短渐尖。果近球形，干燥时黑色。花期春季，果期夏秋季。

Deciduous trees, up to 20 m tall. Bark grayish brown or blackish brown; young branches green, glabrous. Leaves with petiole 25~45 cm or longer, axis slightly flat, grooved adaxially, glabrous or pilosulose; leaflets 5~8 pairs, usually subopposite; petiolule ca. 5 mm; blades adaxially shiny, narrowly elliptic-lanceolate or slightly falcate, thinly papery, base cuneate, slightly asymmetrical, apex acute or shortly acuminate. Fruit subglobose, black when dry. Fl. spring, fr. Summer - autumn.

树干　Trunk
摄影：杨庆松　Photo by: Yang Qingsong

花枝　Flowering branches
摄影：杨庆松　Photo by: Yang Qingsong

果枝　Fruiting branches
摄影：杨庆松　Photo by: Yang Qingsong

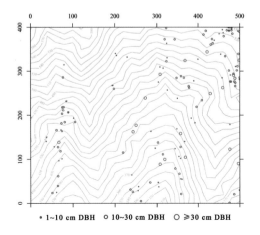

个体分布图　Distribution of individuals

径级分布表 DBH class

胸径区间 (Diameter class) (cm)	个体数 (No. of individuals in the plot)	比例 (Proportion) (%)
1～2	11	7.38
2～5	18	12.08
5～10	38	25.50
10～20	43	28.86
20～30	21	14.09
30～60	18	12.08
≥60	0	0.00

92 红枝柴　　　　　　　　　　　　　　　　　Hóng zhī chái | Oldham Meliosma

Meliosma oldhamii Maxim.
清风藤科 | Sabiaceae

代码 (SpCode) = MELOLD
个体数 (Individual number/20 hm^2) = 63
最大胸径 (Max DBH) = 49.5 cm
重要值排序 (Importance value rank) = 52

落叶乔木，高20 m。奇数羽状复叶，15～30 cm，叶轴、叶柄和小叶两面褐色短柔毛；小叶7～15，薄纸质，基部圆形或宽或狭楔形，边缘疏生有尖锐的锯齿，先端锐尖或急尖。核果球状，红色，直径4～5 mm。花期5月，果期8～9月。

Deciduous trees to 20 m tall. Leaves odd pinnate, 15~30 cm, axis, petiolules and both surfaces of leaflets brownish pubescent; leaflets 7~15, thinly papery, base rounded or broadly or narrowly cuneate, margin sparsely acutely serrate, apex acute or acute-acuminate. Drupe globose, red, 4~5 mm in diam. Fl. May, fr. Aug. - Sep..

枝叶　Branches and leaves
摄影：杨庆松　Photo by: Yang Qingsong

叶背　Leaf abaxial surface
摄影：杨庆松　Photo by: Yang Qingsong

果枝　Fruiting branches
摄影：杨庆松　Photo by: Yang Qingsong

径级分布表 DBH class

胸径等级 (Diameter class) (cm)	个体数 (No. of individuals in the plot)	比例 (Proportion) (%)
1～2	6	9.52
2～5	14	22.22
5～10	9	14.29
10～20	14	22.22
20～30	8	12.70
30～60	12	19.05
≥60	0	0.00

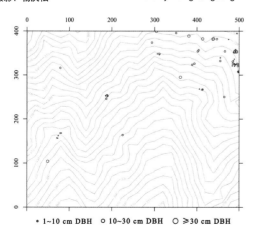

个体分布图　Distribution of individuals

93 笔罗子

Bǐ luó zǐ | Stiffleaf Meliosma

Meliosma rigida Sieb. et Zucc.
清风藤科 | Sabiaceae

代码 (SpCode) = MELRIG
个体数 (Individual number/20 hm^2) = 1
最大胸径 (Max DBH) = 1.2 cm
重要值排序 (Importance value rank) = 153

常绿乔木，高达10 m。芽、幼枝、背面中脉、花序均密被铁锈色柔毛或绒毛。单叶，叶柄1.5～4 cm；叶片倒披针形或狭倒卵形，革质，背面锈色短柔毛，疏生柔毛，或密被绒毛，叶1/3或1/2以下至基部渐狭，边缘有粗锯齿，有时全缘，先端渐尖或尾状渐尖的。核果球状，5～8 mm。花期夏季，果期9～10月。

Evergreen trees to 10 m tall. Buds, young branches, abaxial midveins, and inflorescences ferruginous lanuginous, pilose, or tomentose. Leaves simple; petiole 1.5~4 cm; leaf blade oblanceolate or narrowly obovate, leathery, abaxially ferruginous pubescent, sparsely pilose, or densely tomentose, base attenuate from 1/3 or 1/2 downward to base, margin coarsely serrate, sometimes entire, apex acuminate or caudate-acuminate. Drupe globose, 5~8 mm. Fl. summer, fr. Sep. - Oct..

树干　　Trunk
摄影：杨庆松　Photo by: Yang Qingsong

枝　　Branch
摄影：杨庆松　Photo by: Yang Qingsong

叶　　Leaves
摄影：杨庆松　Photo by: Yang Qingsong

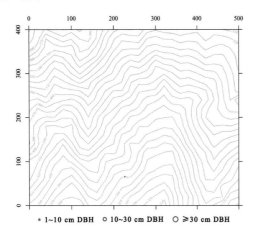

个体分布图 Distribution of individuals

径级分布表 DBH class

胸径区间 (Diameter class) (cm)	个体数 (No. of individuals in the plot)	比例 (Proportion) (%)
1～2	1	100.00
2～5	0	0.00
5～10	0	0.00
10～20	0	0.00
20～30	0	0.00
30～60	0	0.00
≥60	0	0.00

94 枳椇 (拐枣)   Zhǐ jǔ | Turnjujube

Hovenia acerba Lindl.
鼠李科 | Rhamnaceae

代码 (SpCode) = HOVACE
个体数 (Individual number/20 hm²) = 9
最大胸径 (Max DBH) = 55.9 cm
重要值排序 (Importance value rank) = 79

落叶乔木，10～25 m 高。小枝褐色或黑紫色，无毛，具明显的白色皮孔。叶柄2～5 cm，无毛；叶片厚纸质或纸质，无毛或仅下面沿脉疏被短柔毛，基部截形或心形，边缘有细锯齿，很少全缘，先端渐尖。浆果成熟时黄褐色或棕色，近球形；果序轴明显膨大，肉质。花期5月，果期8～10月。

Deciduous trees, 10~25 m tall. Branchlets brown or black-purple, glabrous, with conspicuous white lenticels. Petiole 2~5 cm, glabrous; leaf blade thickly papery to papery, glabrous or pilose on abaxially veins or in vein axils, base truncate or cordate, rarely subrounded or broadly cuneate, margin finely serrulate, rarely subentire, apex acuminate. Fruit yellow-brown or brown at maturity, subglobose; fruiting peduncles and pedicels dilated and fleshy. Fl. May - Jul., fr. Aug. - Oct..

幼苗   Seedling
摄影：刘何铭   Photo by: Liu Heming

花枝   Flowering branches
摄影：杨庆松   Photo by: Yang Qingsong

果枝   Fruiting branches
摄影：杨庆松   Photo by: Yang Qingsong

径级分布表 DBH class

胸径等级 (Diameter class) (cm)	个体数 (No. of individuals in the plot)	比例 (Proportion) (%)
1～2	2	22.22
2～5	0	0.00
5～10	0	0.00
10～20	0	0.00
20～30	1	11.11
30～60	6	66.67
≥60	0	0.00

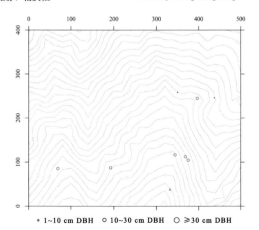

个体分布图 Distribution of individuals

95 长叶冻绿

Cháng yè dòng lǜ | Oriental Buckthorn

Rhamnus crenata Sieb.et Zucc.
鼠李科 | Rhamnaceae

代码 (SpCode) = RHACRE
个体数 (Individual number/20 hm²) = 1
最大胸径 (Max DBH) = 2.8 cm
重要值排序 (Importance value rank) = 150

落叶灌木或小乔木，高可达4 m。幼枝被锈色绒毛。叶柄4-10 (12) mm，密被短柔毛；叶片纸质，背面被短柔毛或至少在脉上有短柔毛，正面无毛，侧脉5～7对，基部楔形或钝，边缘具细圆齿，先端渐尖到尾状渐尖或短急。核果红色,熟时黑色,球形或卵状球。花期5～8月，果期8～10月。

Deciduous shrubs or small trees, up to 4 m tall. Young branchlets ferruginous tomentose. Petiole 4~10(~12) mm, densely pubescent; leaf blade papery, abaxially pubescent or at least pubescent on veins, adaxially glabrous, lateral veins 7~12 pairs, base cuneate or obtuse, margin finely crenate, apex acuminate to caudate-acuminate, or shortly acute. Drupe red, black, or purple-black at maturity, globose or obovoid-globose. Fl. May - Aug., fr. Aug. - Oct..

植株　Whole plant
摄影：杨庆松　Photo by: Yang Qingsong

花枝　Flowering branches
摄影：杨庆松　Photo by: Yang Qingsong

果枝　Fruiting branches
摄影：杨庆松　Photo by: Yang Qingsong

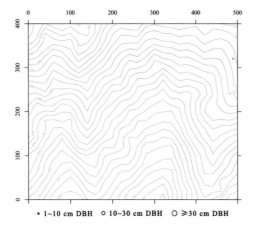
个体分布图　Distribution of individuals

径级分布表 DBH class

胸径区间 (Diameter class) (cm)	个体数 (No. of individuals in the plot)	比例 (Proportion) (%)
1～2	0	0.00
2～5	1	100.00
5～10	0	0.00
10～20	0	0.00
20～30	0	0.00
30～60	0	0.00
≥60	0	0.00

96 中华杜英

Zhōng huá dù yīng | China Elaeocarpus

Elaeocarpus chinensis (Gardn. et Champ.) Hook. f.
杜英科 | Elaeocarpaceae

代码 (SpCode) = ELACHI
个体数 (Individual number/20 hm^2) = 6
最大胸径 (Max DBH) = 5.3 cm
重要值排序 (Importance value rank) = 124

常绿小乔木，高3～7 m。小枝被微柔毛，老时无毛。叶柄1.5～2 cm，纤细，两端稍膨大；叶片卵状披针形或披针形，纸质，背面黑色腺点，基部圆形，很少宽楔形，边缘有细圆齿，先端渐尖。核果椭圆形。花期5～7月，果期6～9月。

Evergreen small trees, 3~7 m tall. Branchlets puberulent, glabrous when old. Petiole 1.5~2 cm, slender, two ends slightly dilated; leaf blade ovate-lanceolate or lanceolate, papery, abaxially black glandular punctate, base rounded, rarely broadly cuneate, margin minutely crenate, apex acuminate. Drupe ellipsoid. Fl. May - Jun., fr. Jun. - Sep..

树干　Trunk
摄影：杨庆松　Photo by: Yang Qingsong

枝叶　Branch and leaves
摄影：杨庆松　Photo by: Yang Qingsong

果枝　Fruiting branches
摄影：严靖　Photo by: Yan Jing

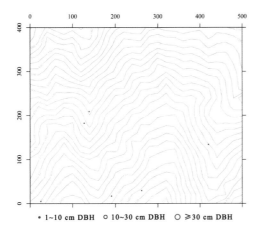

个体分布图　Distribution of individuals

径级分布表　DBH class

胸径等级 (Diameter class) (cm)	个体数 (No. of individuals in the plot)	比例 (Proportion) (%)
1～2	2	33.33
2～5	3	50.00
5～10	1	16.67
10～20	0	0.00
20～30	0	0.00
30～60	0	0.00
≥60	0	0.00

97 杜英

Dù yīng | Common Elaeocarpus

Elaeocarpus decipiens Hemsley
杜英科 | Elaeocarpaceae

代码 (SpCode) = ELADEC
个体数 (Individual number/20 hm^2) = 9
最大胸径 (Max DBH) = 13.2 cm
重要值排序 (Importance value rank) = 117

常绿乔木，5～15 m高。叶柄1～2 cm，上端不膨大；叶片革质或厚纸质，无毛，侧脉7～9每侧，背面突起，正面不明显，基部下延，边缘有细锯齿，先端渐尖，尖钝头。核果椭圆形，2～3.5×1.5～2 cm。花期6～7月，果期11月～翌年1月。

Evergreen trees, 5~15 m tall. Petiole 1~2 cm, not swollen at upper end; leaf blade leathery or thickly papery, glabrous, lateral veins 7~9 per side, prominent abaxially, inconspicuous adaxially, base attenuate, margin serrulate, apex acuminate, acumen obtuse. Drupe ellipsoid, 2~3.5 × 1.5~2 cm. Fl. Jun. - Jul., fr. Nov. - Jan. of following year.

树干　　Trunk
摄影：杨庆松　Photo by: Yang Qingsong

枝叶　　Branch and leaves
摄影：杨庆松　Photo by: Yang Qingsong

果　　Fruits
摄影：杨庆松　Photo by: Yang Qingsong

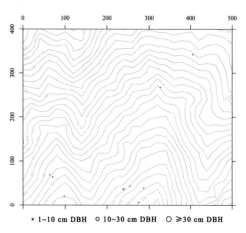

个体分布图 Distribution of individuals

径级分布表 DBH class

胸径区间 (Diameter class) (cm)	个体数 (No. of individuals in the plot)	比例 (Proportion) (%)
1～2	4	44.44
2～5	4	44.44
5～10	0	0.00
10～20	1	11.11
20～30	0	0.00
30～60	0	0.00
≥60	0	0.00

98 南京椴

Nán jīng duàn | Nanjing Linden

Tilia miqueliana Maxim.
椴树科 | Tiliaceae

代码 (SpCode) = TILMIQ
个体数 (Individual number/20 hm^2) = 5
最大胸径 (Max DBH) = 37.9 cm
重要值排序 (Importance value rank) = 113

落叶乔木，10 (20) m高。树皮灰白色；小枝和顶芽被黄褐色绒毛。叶柄3～4 cm，被星状绒毛；叶片卵圆形，背面灰色或灰黄色星状绒毛，正面无毛，基部心形，很少偏斜，边缘有锯齿，先端锐尖。聚伞花序3～12花，长6～8 cm。果球状，不具棱。花期7月，果期8～10月。

Deciduous trees, 10(~20) m tall. Bark gray-white; branchlets and terminal bud yellow-brown tomentose. Petiole 3~4 cm, stellate tomentose; leaf blade ovate-orbicular, abaxially gray or gray-yellow stellate tomentose, adaxially glabrous, base cordate, rarely oblique, margin serrate, apex acute. Cymes 3~12-flowered, 6~8 cm. Fruit globose, not angled. Fl. Jul., Fr. Aug. - Oct..

树干　　Trunk
摄影：杨庆松　　Photo by: Yang Qingsong

枝叶　　Branch and leaves
摄影：杨庆松　　Photo by: Yang Qingsong

花枝　　Flowering branches
摄影：汪远　　Photo by: Wang Yuan

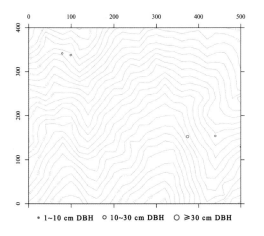

个体分布图　Distribution of individuals

径级分布表　DBH class

胸径等级 (Diameter class) (cm)	个体数 (No. of individuals in the plot)	比例 (Proportion) (%)
1～2	0	0.00
2～5	0	0.00
5～10	1	20.00
10～20	2	40.00
20～30	1	20.00
30～60	1	20.00
≥60	0	0.00

99 毛花连蕊茶

Máo huā lián ruǐ chá | Hairstalk Tea

Camellia fraterna Hance
山茶科 | Theaceae

代码 (SpCode) = CAMFRA
个体数 (Individual number/20 hm^2) = 9263
最大胸径 (Max DBH) = 12.4 cm
重要值排序 (Importance value rank) = 6

常绿灌木，1~4 m高。幼枝密被开展长柔毛和丝毛。叶柄3~5 mm，具长柔毛；叶片薄革质，叶面深绿色，有光泽，中脉被微糙硬毛，基部楔形或宽楔形，边缘有细锯齿，先端钝的渐尖。花腋生，单生或成对，花冠白色，直径2.5~4 cm。蒴果球形，直径1.5~2 cm。花期2~3月，果期9~10月。

Evergreen shrubs, 1~4 m tall. Young branches densely spreading villous and hirsute. Petiole 3~5 mm, villous; leaf blade thinly leathery, adaxially dark green, shiny, and hirtellous along midvein, base cuneate to broadly cuneate, margin serrulate, apex bluntly acuminate. Flowers axillary, solitary or paired, white or sometimes pale pink, 2.5~4 cm in diam. Capsule globose, 1.5~2 cm in diam. Fl. Feb. - Mar., fr. Sep. - Oct..

枝叶 Branch and leaves
摄影：杨庆松 Photo by: Yang Qingsong

花枝 Flowering branch
摄影：杨庆松 Photo by: Yang Qingsong

果 Fruit
摄影：杨庆松 Photo by: Yang Qingsong

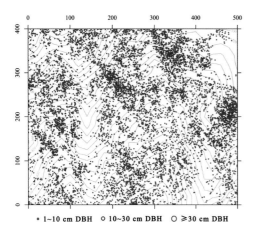
个体分布图 Distribution of individuals

径级分布表 DBH class

胸径区间 (Diameter class) (cm)	个体数 (No. of individuals in the plot)	比例 (Proportion) (%)
1~2	4191	45.24
2~5	4604	49.70
5~10	459	4.96
10~20	9	0.10
20~30	0	0.00
30~60	0	0.00
≥60	0	0.00

100 山茶

Shān chá | Japan Camellia

Camellia japonica Linn.
山茶科 | Theaceae

代码 (SpCode) = CAMJAP
个体数 (Individual number/20 hm^2) = 66
最大胸径 (Max DBH) = 22.3 cm
重要值排序 (Importance value rank) = 91

常绿灌木或乔木，1.5～6 (11) m高。幼枝略带紫色棕色，无毛。叶柄5～10 mm，无毛或短柔毛；叶片革质，背面浅绿色和棕色腺点，叶面深绿色，两面无毛，较厚，基部楔形或宽楔形，边缘有细锯齿，先端短渐尖，具钝尖。花腋生或近顶生，单生或成对，红色，直径6～10 cm，近无柄。蒴果球状。花期1～3月，果期9～10月。

Evergreen shrubs or trees, 1.5~6(~11) m tall. Young branches purplish brown, glabrous. Petiole 5~10 mm, glabrous or adaxially pubescent; leaf blade leathery, abaxially pale green and brown glandular punctate, adaxially dark green, both surfaces glabrous, thick, base cuneate to broadly cuneate, margin serrulate, apex shortly acuminate and with an obtuse tip. Flowers axillary or subterminal, solitary or paired, red, 6~10 cm in diam., subsessile. Capsule globose. Fl. Jan.-Mar., fr. Sep. - Oct..

树干　Trunk
摄影：杨庆松　Photo by: Yang Qingsong

枝叶　Branch and leaves
摄影：杨庆松　Photo by: Yang Qingsong

花　Flower
摄影：汪远　Photo by: Wang Yuan

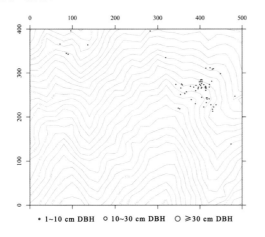

个体分布图　Distribution of individuals

径级分布表　DBH class

胸径等级 (Diameter class) (cm)	个体数 (No. of individuals in the plot)	比例 (Proportion) (%)
1～2	26	39.39
2～5	35	53.03
5～10	4	6.06
10～20	0	0.00
20～30	1	1.52
30～60	0	0.00
≥60	0	0.00

101 茶 Chá | Tea

Camellia sinensis (Linn.) Kuntze
山茶科 | Theaceae

代码 (SpCode) = CAMSIN
个体数 (Individual number/20 hm^2) = 6
最大胸径 (Max DBH) = 1.7 cm
重要值排序 (Importance value rank) = 126

常绿灌木或乔木，1~5 (9) m高。幼枝淡灰黄色，无毛；当年小枝紫红色，被白色短柔毛；顶芽银灰色绢毛。叶柄4~7 mm；叶片革质，背面淡绿色，无毛或被短柔毛，正面深绿色，有光泽，无毛，两面可见网状脉，基部楔形或宽楔形，边缘有锯齿的到有细锯齿。花1~3朵腋生，白色。蒴果扁球形。花期10月至翌年2月，果期翌年8~10月。

Evergreen shrubs or trees, 1~5(~9) m tall. Young branches grayish yellow, glabrous; current year branchlets purplish red, white pubescent; terminal buds silvery gray sericeous. Petiole 4~7 mm; leaf blade leathery, abaxially pale green and glabrous or pubescent, adaxially dark green, shiny, and glabrous, reticulate veins visible on both surfaces, base cuneate to broadly cuneate, margin serrate to serrulate. Flowers 1~3 at leaf axil, white. Capsule oblate. Fl. Oct. - Feb. of following year, fr. Aug. - Oct. of following year.

枝叶　Branch and leaves
摄影：杨庆松　Photo by: Yang Qingsong

叶　Leaves
摄影：杨庆松　Photo by: Yang Qingsong

花　Flower
摄影：杨庆松　Photo by: Yang Qingsong

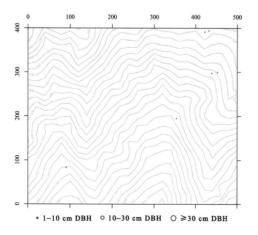

个体分布图　Distribution of individuals

径级分布表　DBH class

胸径区间 (Diameter class) (cm)	个体数 (No. of individuals in the plot)	比例 (Proportion) (%)
1~2	6	100.00
2~5	0	0.00
5~10	0	0.00
10~20	0	0.00
20~30	0	0.00
30~60	0	0.00
≥60	0	0.00

102 红淡比 (杨桐)

Hóng dàn bǐ | Japan Cleyera

Cleyera japonica Thunb.
山茶科 | Theaceae

代码 (SpCode) = CLEJAP
个体数 (Individual number/20 hm^2) = 2379
最大胸径 (Max DBH) = 31.0 cm
重要值排序 (Importance value rank) = 11

常绿灌木或乔木，2～10 m高。幼枝淡灰棕色，圆柱状；顶芽长圆锥形，1～1.5 cm，无毛。叶柄7～10 (12) mm，无毛；叶片厚革质，背面淡绿色，叶面深绿色，有光泽，两面无毛，中脉正面稍凹下，边缘全缘。花2～4朵腋生，花瓣白色。果成熟时略带紫黑色，球状。花期5～7月，果期10～12月。

Evergreen shrubs or trees, 2~10 m tall. Young branches grayish brown, terete; terminal buds long conic, 1~1.5 cm, glabrous. Petiole 7~10 (~12) mm, glabrous; leaf blade thick leathery, abaxially pale green, adaxially dark green and shiny, both surfaces glabrous, midvein level adaxially slightly impressed, margin entire. Flowers 2~3 clusterd at leaf axil, petals white. Fruit purplish black when mature, globose. Fl. May - Jul., fr. Oct. - Dec..

花 Flowers
摄影：杨庆松 Photo by: Yang Qingsong

叶 Leaves
摄影：杨庆松 Photo by: Yang Qingsong

果枝 Fruiting branch
摄影：杨庆松 Photo by: Yang Qingsong

径级分布表 DBH class

胸径等级 (Diameter class) (cm)	个体数 (No. of individuals in the plot)	比例 (Proportion) (%)
1～2	655	27.53
2～5	682	28.67
5～10	398	16.73
10～20	589	24.76
20～30	52	2.19
30～60	3	0.13
≥60	0	0.00

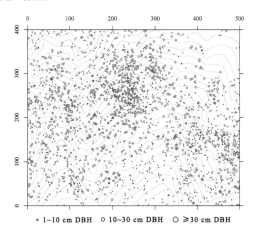

个体分布图 Distribution of individuals

103 细枝柃

Xì zhī líng | Slenderbranch Eurya

Eurya loquaiana Dunn
山茶科 | Theaceae

代码 (SpCode) = EURLOQ
个体数 (Individual number/20 hm^2) = 20257
最大胸径 (Max DBH) = 16.4 cm
重要值排序 (Importance value rank) = 1

常绿灌木或小乔木，2～10 m高。当年小枝黄绿色，圆柱状，纤细，被微柔毛。叶柄2～4 mm，被微柔毛；叶片薄革质或纸质。花1～4朵簇生叶腋，花冠白色。果成熟时黑色，球状，直径3～4 mm。花期10～12月，果期翌年7～9月。

Evergreen shrubs or trees, 2~10 m tall. Current year branchlets yellowish green, terete, slender, puberulent; terminal buds puberulent and pubescent. Petiole 2~4 mm, puberulent; leaf blade thinly leathery to papery. Flowers 1~4 clusterd at leaf axil, petals white. Fruit black when mature, globose, 3~4 mm in diam. Fl. Oct. - Dec., fr. Jul. - Sep.. of following year.

叶背　Leaf abaxial surface
摄影：杨庆松　Photo by: Yang Qingsong

花枝　Flowering branch
摄影：雷霄　Photo by: Lei Xiao

果枝　Fruiting branch
摄影：杨庆松　Photo by: Yang Qingsong

径级分布表 DBH class

胸径区间 (Diameter class) (cm)	个体数 (No. of individuals in the plot)	比例 (Proportion) (%)
1～2	7152	35.31
2～5	12001	59.24
5～10	1072	5.29
10～20	32	0.16
20～30	0	0.00
30～60	0	0.00
≥60	0	0.00

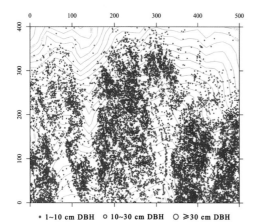

个体分布图 Distribution of individuals

104 格药柃

Gé yào líng | Muricate Eurya

Eurya muricata Dunn
山茶科 | Theaceae

代码 (SpCode) = EURMUR
个体数 (Individual number/20 hm^2) = 760
最大胸径 (Max DBH) = 13.3 cm
重要值排序 (Importance value rank) = 36

常绿灌木或小乔木，2~6 m高。当年小枝黄绿色，圆柱状，粗壮，无毛或疏生短柔毛。叶柄4~5 mm；叶片长圆状椭圆形至倒椭圆形，革质，背面淡绿色到黄绿色，叶面深绿色，有光泽，边缘有细锯齿。花单性，雌雄异株。花1~5朵簇生于叶腋。果成熟时略带紫黑色，球状，直径4~5 mm。花期9~12月，果期翌年6~9月。

Evergreen shrubs or small trees, 2~6 m tall. Current year branchlets yellowish green, terete, strong, glabrous or sparsely pubescent. Petiole 4~5 mm; leaf blade oblong-elliptic to elliptic, leathery, abaxially pale green to yellowish green, adaxially dark green and shiny, margin serrulate. Flowers unisexual, dioecious. Fruit purplish black when mature, globose, 4~5 mm in diam. Fl. Sep. - Dec., fr. Jun. - Sep. of following year.

花枝　Flowering branch
摄影：杨庆松　Photo by: Yang Qingsong

枝叶　Branch and leaves
摄影：杨庆松　Photo by: Yang Qingsong

果枝　Fruiting branch
摄影：杨海波　Photo by: Yang Haibo

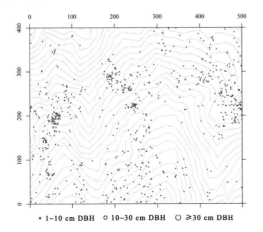

个体分布图　Distribution of individuals

径级分布表 DBH class

胸径等级 (Diameter class) (cm)	个体数 (No. of individuals in the plot)	比例 (Proportion) (%)
1~2	280	36.84
2~5	410	53.95
5~10	62	8.16
10~20	8	1.05
20~30	0	0.00
30~60	0	0.00
≥60	0	0.00

105 窄基红褐栲

Zhǎi jī hóng hè líng | Attenuate Eurya

Eurya rubiginosa var. *attenuata* H.T. Chang
山茶科 | Theaceae

代码 (SpCode) = EURRUB

个体数 (Individual number/20 hm^2) = 2036

最大胸径 (Max DBH) = 13.1 cm

重要值排序 (Importance value rank) = 23

常绿灌木2.5～3.5 m。老枝灰白色；幼枝淡灰棕色，具两棱。叶柄长约2 mm或更少；叶片长圆状披针形，革质，干后背面红棕色，正面深绿色，两面无毛，基部楔形或宽楔形，先端锐尖至短渐尖，钝尖，边缘有细锯齿，反卷紧密。花单生或2～3朵簇生叶腋。果成熟时略带紫黑色，球状到卵球形，长约4 mm。花期10～11月，果期翌年4～8月。

Evergreen shrubs 2.5~3.5 m tall. Older branches grayish white; young branches grayish brown, 2-ribbed. Petiole ca. 2 mm or less; leaf blade oblong-lanceolate, leathery, abaxially reddish brown when dry, adaxially dark green, both surfaces glabrous, base cuneate or broadly cuneate, apex acute to shortly acuminate and with an obtuse tip, margin closely serrulate and revolute. Flowers solitary or 2~3 clustered at leaf axil. Fruit purplish black when mature, globose to ovoid, ca. 4 mm. Fl. Oct. - Nov., fr. Apr. - Aug. of following year.

花枝　Flowering branch
摄影：杨庆松　Photo by: Yang Qingsong

枝叶　Branch and leaves
摄影：杨庆松　Photo by: Yang Qingsong

果枝　Fruiting branch
摄影：杨庆松　Photo by: Yang Qingsong

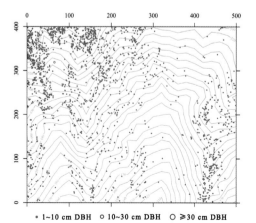
个体分布图 Distribution of individuals

径级分布表 DBH class

胸径区间 (Diameter class) (cm)	个体数 (No. of individuals in the plot)	比例 (Proportion) (%)
1～2	1230	60.41
2～5	778	38.21
5～10	27	1.33
10～20	1	0.05
20～30	0	0.00
30～60	0	0.00
≥60	0	0.00

106 木荷

Mù hé | Gugertree

Schima superba Gardn. et Champ.
山茶科 | Theaceae

代码 (SpCode) = SCHSUP
个体数 (Individual number/20 hm^2) = 1233
最大胸径 (Max DBH) = 68.9 cm
重要值排序 (Importance value rank) = 8

常绿乔木5~20 m高。叶柄1~2 cm；叶片椭圆形至长圆状椭圆形，7~13×2.5~4 (6) cm，薄革质至革质，边缘波状钝齿从基部1/2顶部，先端渐尖。花4~8朵总状花序，2~3 cm直径。花瓣白色，倒卵形，1~1.5 cm。蒴果近球形，直径1~2 cm。花期6~8月，果期10~12月。

Evergreen trees 5~20 m tall. Petiole 1~2 cm; leaf blade elliptic to oblong-elliptic, 7~13 × 2.5~4(~6) cm, thinly leathery to leathery, base cuneate, margin undulately obtusely crenate from basal 1/2 apically, apex acuminate. Flowers 4-8 in a raceme, 2-3 cm in diam. Petals white, obovate, 1~1.5 cm. Capsule subglobose, 1~2 cm in diam. Fl. Jun. - Aug., fr. Oct. - Dec..

枝叶　Branch and leaves
摄影：杨庆松　Photo by: Yang Qingsong

花　Flowers
摄影：杨庆松　Photo by: Yang Qingsong

果　Fruit
摄影：杨庆松　Photo by: Yang Qingsong

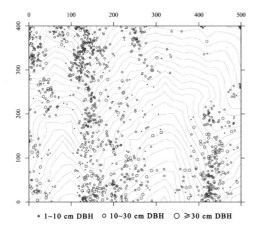

个体分布图　Distribution of individuals

径级分布表　DBH class

胸径等级 (Diameter class) (cm)	个体数 (No. of individuals in the plot)	比例 (Proportion) (%)
1~2	39	3.16
2~5	109	8.84
5~10	216	17.52
10~20	402	32.60
20~30	307	24.90
30~60	157	12.73
≥60	3	0.24

107 厚皮香

Hòu pí xiāng | Ternstroemia

Ternstroemia gymnanthera (Wight et Arn.) Beddome
山茶科 | Theaceae

代码 (SpCode) = TERGYM
个体数 (Individual number/20 hm^2) = 123
最大胸径 (Max DBH) = 22.9 cm
重要值排序 (Importance value rank) = 61

常绿小乔木或灌木，1.5~10 m高。全株无毛。叶革质或薄革质，通常聚生于枝端，呈假轮生状，椭圆形、椭圆状倒卵形至长圆状倒卵形，叶面深绿色，有光泽，基部楔形，边缘全缘或具疏锯齿，先端锐尖至短渐尖，具钝尖。成熟时的果紫红色，球状。花期5~7月，果期8~10月。

Evergreen small tree or shrubs, 1.5~10 m tall. Plant glabrous. Petiole 0.7~1.3 cm, adaxially grooved; leaf blade obovate, oblong-obovate, or broadly elliptic, leathery, abaxially pale green becoming reddish brown when dry, adaxially dark green and shiny, base cuneate, margin entire or apically sparsely serrate, apex acute to shortly acuminate and with an obtuse tip. Fruit purplish red when mature, globose. Fl. May - Jul., fr. Aug. - Oct..

花　　Flowers
摄影：朱鑫鑫　　Photo by: Zhu Xinxin

叶　　Leaves
摄影：杨庆松　　Photo by: Yang Qingsong

果枝　　Fruiting branches
摄影：汪远　　Photo by: Wang Yuan

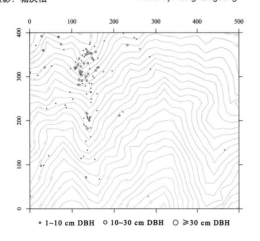

个体分布图　Distribution of individuals

径级分布表 DBH class

胸径区间 (Diameter class) (cm)	个体数 (No. of individuals in the plot)	比例 (Proportion) (%)
1~2	30	24.39
2~5	34	27.64
5~10	19	15.45
10~20	38	30.89
20~30	2	1.63
30~60	0	0.00
≥60	0	0.00

108 山桐子　　Shān tóng zǐ | Manyfruit Idesia

Idesia polycarpa Maxim.
大风子科 | Flacourtiaceae

代码 (SpCode) = IDEPOL
个体数 (Individual number/20 hm^2) = 24
最大胸径 (Max DBH) = 33.4 cm
重要值排序 (Importance value rank) = 92

落叶乔木，高可达21 m；树皮浅灰色，不剥落。叶柄带红色，通常长5～15 cm或更多，无毛，基部稍膨大；叶片宽卵形，薄革质，下面有白粉，叶片通常基出5 (7) 脉，基部心形，边缘有粗锯齿，先端逐渐或突然渐尖。浆果成熟期紫红色，扁圆形。种子红棕色，宽卵形。花期4～5月，果期10～11月。

Deciduous trees, up to 21 m tall; bark grayish, not flaking. Petiole reddish, usually long, 5~15 cm or more, glabrous, base slightly dilated; leaf blade broadly ovate, thinly leathery, blade usually 5(~7)-veined from base, base cordate, margin serrate, apex gradually or more abruptly acuminate. Berry purple-red or orangered when mature. Seeds drying reddish brown, broadly ovoid. Fl. Apr. - May, fr. Oct. - Nov..

树干　　Trunk
摄影：杨庆松　　Photo by: Yang Qingsong

枝叶　　Branch and leaves
摄影：杨庆松　　Photo by: Yang Qingsong

果枝　　Fruiting branches
摄影：汪远　　Photo by: Wang Yuan

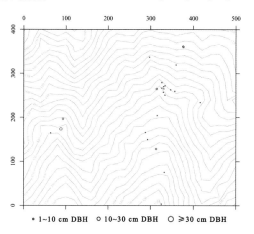

个体分布图 Distribution of individuals

径级分布表 DBH class

胸径等级 (Diameter class) (cm)	个体数 (No. of individuals in the plot)	比例 (Proportion) (%)
1～2	3	12.50
2～5	12	50.00
5～10	3	12.50
10～20	2	8.33
20～30	3	12.50
30～60	1	4.17
≥60	0	0.00

109 胡颓子

Hú tuí zǐ | Thorny Elaeagnus

Elaeagnus pungens Thunb.
胡颓子科 | Elaeagnaceae

代码 (SpCode) = ELAPUN

个体数 (Individual number/20 hm^2) = 21

最大胸径 (Max DBH) = 4.5 cm

重要值排序 (Importance value rank) = 107

常绿灌木，3~4 m高。常具刺。叶柄粗壮，5~15 mm，具皱纹，棕色鳞片；叶片长圆形至狭卵形，5~10 × 1.8~3.5 cm，革质，背面密被白色、通常棕色鳞片，正面无毛，有光泽，基部圆形，边缘具齿，常微反卷呈明显的波状边缘。花1~3朵生于叶腋小枝上，银白色，下垂。核果长圆形，1.2~1.5 cm，被棕色鳞片。花期9~12月，果期翌年4~6月。

Evergreen shrubs, 3~4 m tall. Spines frequent. Petiole robust, 5~15 mm, rugose, brown scaly; leaf blade oblong to narrowly so, 5~10 × 1.8~3.5 cm, leathery, abaxially with dense whitish and usually also brown scales, adaxially glabrous and glossy, base rounded, margin obsoletely toothed with prominently undulate margins. Flowers 1~3 clusterd on branchlet in leaf axil, silvery white, drooping. Drupe oblong, 1.2~1.5 cm, brown scaly. Fl. Sep. - Dec., fr. Apr. - Jun. of following year.

叶背　Leaf abaxial surface
摄影：杨庆松　Photo by: Yang Qingsong

叶　Leaves
摄影：杨庆松　Photo by: Yang Qingsong

花枝　Flowering branches
摄影：杨庆松　Photo by: Yang Qingsong

径级分布表 DBH class

胸径区间 (Diameter class) (cm)	个体数 (No. of individuals in the plot)	比例 (Proportion) (%)
1~2	8	38.10
2~5	13	61.90
5~10	0	0.00
10~20	0	0.00
20~30	0	0.00
30~60	0	0.00
≥60	0	0.00

个体分布图 Distribution of individuals

110 毛八角枫

Máo bā jiǎo fēng | Kurz Alangium

Alangium kurzii Craib
八角枫科 | Alangiaceae

代码 (SpCode) = ALAKUR
个体数 (Individual number/20 hm^2) = 51
最大胸径 (Max DBH) = 19.7 cm
重要值排序 (Importance value rank) = 86

落叶小乔木或灌木，5～10 m高。小枝平滑，疏生短柔毛，后脱落。叶柄2.5～4 cm；叶片近圆形或宽卵形，12～14×7～9 cm，纸质，背面被短柔毛，正面无毛，基部圆形或近圆形，边缘全缘，先端渐尖。聚伞花序5～7花；花瓣白色或米色。核果熟时深紫色近黑色，椭圆形。花期5～6月，果期9月。

Deciduous trees or shrubs, to 5~10 m tall. Twigs smooth, sparsely pubescent, glabrescent. Petiole 2.5~4 cm; leaf blade suborbicular or broadly ovate, 12~14 × 7~9 cm, papery, abaxially pubescent, adaxially glabrous, base rounded or subrounded, margin entire, apex acuminate. Inflorescences 5~7-flowered. Petals white or cream-colored. Mature drupe dark violet to nearly black, elliptic, 1.2~1.5 mm. Fl. May - Jun., fr. Sep..

花 Flower
摄影：杨庆松 Photo by: Yang Qingsong

枝叶 Branch and leaves
摄影：杨庆松 Photo by: Yang Qingsong

果枝 Fruiting branch
摄影：杨庆松 Photo by: Yang Qingsong

径级分布表 DBH class

胸径等级 (Diameter class) (cm)	个体数 (No. of individuals in the plot)	比例 (Proportion) (%)
1~2	13	25.49
2~5	7	13.73
5~10	14	27.45
10~20	17	33.33
20~30	0	0.00
30~60	0	0.00
≥60	0	0.00

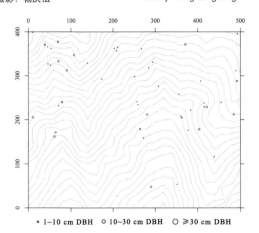

个体分布图 Distribution of individuals

111 赤楠

Chì nán | Boxleaf Syzygium

Syzygium buxifolium Hook. et Arn.
桃金娘科 | Myrtaceae

代码 (SpCode) = SYZBUX
个体数 (Individual number/20 hm^2) = 461
最大胸径 (Max DBH) = 12.4 cm
重要值排序 (Importance value rank) = 42

常绿灌木。小枝有棱，干后黑褐色。叶柄长约2 mm；叶片革质，1～3×0.5～2 (2.2) cm，干后背面稍苍白，正面暗褐色，没有光泽，基部宽楔形或楔形，先端圆或钝，有时具钝尖头。聚伞花序顶生，花白色。果红色略带紫黑色，球形。花期6～8月，果期10～12月。

Evergreen shrubs. Branchlets blackish brown when dry, 4~ or 6-angled. Petiole ca. 2 mm; leaf blade leathery, 1~3 × 0.5~2(~2.2) cm, abaxially slightly pale when dry, adaxially dark brown and not glossy when dry, base broadly cuneate or cuneate, apex rounded, obtuse, or acute and sometimes with an obtuse cusp. Inflorescences terminal, cymes, flowers white. Fruit red turning purplish black, globose. Fl. Jun. - Aug., fr. Oct. - Dec..

花枝　Flowering branch
摄影：杨庆松　Photo by: Yang Qingsong

枝叶　Branch and leaves
摄影：杨庆松　Photo by: Yang Qingsong

果枝　Fruiting branches
摄影：杨庆松　Photo by: Yang Qingsong

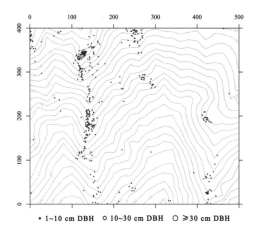

个体分布图　Distribution of individuals

径级分布表　DBH class

胸径区间 (Diameter class) (cm)	个体数 (No. of individuals in the plot)	比例 (Proportion) (%)
1～2	152	32.97
2～5	224	48.59
5～10	80	17.35
10～20	5	1.08
20～30	0	0.00
30～60	0	0.00
≥60	0	0.00

112 黄毛楤木

Huáng máo cōng mù | Yellowhair Aralia

Aralia chinensis Linn.
五加科 | Araliaceae

代码 (SpCode) = ARACHI
个体数 (Individual number/20 hm²) = 2
最大胸径 (Max DBH) = 4.9 cm
重要值排序 (Importance value rank) = 139

落叶灌木，2~5 (8) m高。树皮灰色，疏生粗壮直刺。小枝疏生细刺。叶二或三回羽状复叶；叶柄长约50 cm，具刺；小叶柄3~5 mm；每羽片有小叶5~11 (13)，小叶纸质，边缘有细锯齿，先端渐尖。果球状，直径约3~4 mm。花期7~9月，果期9~12月。

Deciduous shrubs, 2~5(~8) m tall. Bark grey with rough prickles. Branches armed with sparse prickles. Leaves 2(or 3)-pinnately compound; petiole to ca. 50 cm, prickly; petiolules 3~5 mm; leaflets 5~11 (~13) per pinna, papery, margin serrulate, apex acuminate. Fruit globose, ca. 3~4 mm in diam. Fl. Jul. - Sep., fr. Sep. - Dec..

茎　Stem
摄影：杨庆松　Photo by: Yang Qingsong

花枝　Flowering branch
摄影：葛斌杰　Photo by: Ge Binjie

叶　Leaves
摄影：杨庆松　Photo by: Yang Qingsong

径级分布表 DBH class

胸径等级 (Diameter class) (cm)	个体数 (No. of individuals in the plot)	比例 (Proportion) (%)
1~2	0	0.00
2~5	2	100.00
5~10	0	0.00
10~20	0	0.00
20~30	0	0.00
30~60	0	0.00
≥60	0	0.00

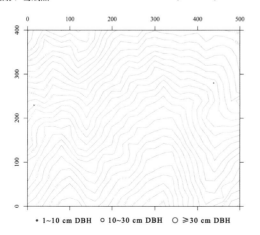

• 1~10 cm DBH　○ 10~30 cm DBH　○ ≥30 cm DBH

个体分布图 Distribution of individuals

113 棘茎楤木

Jí jīng cōng mù | Spinystem Aralia

Aralia echinocaulis Hand.-Mazz.
五加科 | Araliaceae

代码 (SpCode) = ARAECH
个体数 (Individual number/20 hm^2) = 4
最大胸径 (Max DBH) = 1.8 cm
重要值排序 (Importance value rank) = 131

落叶灌木或小乔木，2~10 m高。枝密生褐色细长直刺，刺长5~20 mm。叶为二回羽状复叶，叶柄25~40 cm，通常无刺；每羽片有小叶5~9，膜质或纸质，两面无毛，背面有白霜，基部圆形至钝，边缘有锯齿，先端渐尖。果实球形至近球形，直径3.5~4.5 mm。花期6~8月，果期9~11月。

Deciduous shrubs or small trees, 2~10 m tall. Branches with dense, brownish, slender needlelike prickles 5~20 mm. Leaves 2-pinnately compound, petiole 25~40 cm; leaflets 5~9 per pinna, membranous to papery, both surfaces glabrous, abaxially glaucous, base rounded to obtuse, margin serrate, apex acuminate. Fruit globose to subglobose, 3.5~4.5 mm in diam. Fl. Jun. - Aug., fr. Sep. - Nov..

植株 Plant
摄影：杨庆松 Photo by: Yang Qingsong

茎 Stem
摄影：杨庆松 Photo by: Yang Qingsong

叶 Leaves
摄影：杨庆松 Photo by: Yang Qingsong

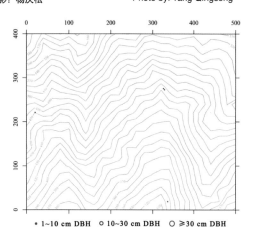
个体分布图 Distribution of individuals

径级分布表 DBH class

胸径区间 (Diameter class) (cm)	个体数 (No. of individuals in the plot)	比例 (Proportion) (%)
1~2	4	100.00
2~5	0	0.00
5~10	0	0.00
10~20	0	0.00
20~30	0	0.00
30~60	0	0.00
≥60	0	0.00

114 刺楸 | Cì qiū | Septemlobate Kalopanax

Kalopanax septemlobus (Thunb.) Koidz.
五加科 | Araliaceae

代码 (SpCode) = KALSEP
个体数 (Individual number/20 hm^2) = 14
最大胸径 (Max DBH) = 63.0 cm
重要值排序 (Importance value rank) = 72

落叶乔木，高达30 m。树皮暗灰棕色；枝粗壮，散生粗刺。叶柄无毛，8~50 cm；叶片纸质，在长枝上互生，短枝上簇生，近圆形，9~25 (~35) cm，背面暗绿色，无毛或近无毛，通常幼时稍具短柔毛，5~7裂；基部心形或圆形至近截形，边缘有锯齿，先端渐尖。果实成熟时暗蓝色，直径3~5 mm。花期7~8月，果期9~10月。

Deciduous trees, to 30 m tall. Bark dark grey-brown; branches stout, with numerous prickles. Petiole glabrous, 8~50 cm; leaf blade papery, alternated in long branch, tufted in short branch, suborbicular, papery, 9~25(~35) cm wide, abaxially dark green and glabrous or nearly so, adaxially light green and usually slightly pubescent when young, 5~7-lobed; base cordate or rounded to nearly truncate, margin serrate, apex acuminate. Fruit dark blue at maturity, 3~5 mm in diam. Fl. Jul. - Aug., fr. Sep. - Oct..

皮刺　　Prickles
摄影：杨庆松　　Photo by: Yang Qingsong

幼苗和花序　Seedling and Inflorescence
摄影：杨庆松、徐晔春　Photo by: Yang Qingsong, Xu Yechun

叶　　Leaf
摄影：杨庆松　　Photo by: Yang Qingsong

径级分布表 DBH class

胸径等级 (Diameter class) (cm)	个体数 (No. of individuals in the plot)	比例 (Proportion) (%)
1~2	0	0.00
2~5	2	14.29
5~10	0	0.00
10~20	0	0.00
20~30	3	21.43
30~60	8	57.14
≥60	1	7.14

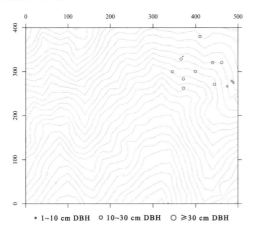

个体分布图 Distribution of individuals

115 灯台树

Dēng tái shù | Lampstandtree

Cornus controversa Hemsl.
山茱萸科 | Cornaceae

代码 (SpCode) = CORCON

个体数 (Individual number/20 hm^2) = 11

最大胸径 (Max DBH) = 21.8 cm

重要值排序 (Importance value rank) = 102

落叶乔木，高13～20 m高。树皮暗灰色或淡黄灰色，平滑。叶互生，纸质，宽卵形或宽椭圆状卵形，5～13×3～9 cm，背面淡或灰绿色，疏生短柔毛具贴伏毛，具乳突，叶脉6或7 (9)，背面凸起，略带紫色，基部近圆形，先端锐尖或渐尖。伞房状聚伞花序顶生。果紫红色或青黑色，球状。花期5～6月，果期7～9月。

Deciduous trees, 13~20 m tall. Bark dark gray or yellowish gray, smooth. Leaf blade alternate, papery, broadly ovate or broadly elliptic-ovate, 5~13 × 3~9 cm, abaxially light or grayish green, sparsely pubescent with appressed trichomes, papillate, veins 6 or 7 (~9), abaxially raised and slightly purplish, base subrounded, apex acute or acuminate. Corymbose cymes terminal. Fruit purplish red or bluish black, globose. Fl. May - Jun., fr. Jul. - Sep..

树干　Trunk
摄影：杨庆松　Photo by: Yang Qingsong

枝叶　Branch and leaves
摄影：杨庆松　Photo by: Yang Qingsong

花枝　Flowering branch
摄影：杨庆松　Photo by: Yang Qingsong

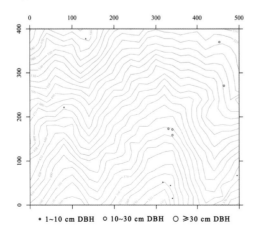
个体分布图　Distribution of individuals

径级分布表　DBH class

胸径区间 (Diameter class) (cm)	个体数 (No. of individuals in the plot)	比例 (Proportion) (%)
1～2	1	9.09
2～5	1	9.09
5～10	4	36.36
10～20	2	18.18
20～30	3	27.27
30～60	0	0.00
≥60	0	0.00

116 四照花

Sì zhào huā | Chinese Kousa Dogwood

Cornus kousa subsp. *chinensis* (Osborn) Q. Y. Xiang
山茱萸科 | Cornaceae

代码 (SpCode) = CORKOU
个体数 (Individual number/20 hm^2) = 83
最大胸径 (Max DBH) = 33.6 cm
重要值排序 (Importance value rank) = 56

落叶乔木，3~10 m高。树皮灰褐色，光滑。叶对生，背面通常密被乳突和短柔毛具贴伏毛，通常具一簇白色或棕色的柔长毛脉腋，脉4或5对，基部楔形到圆形，先端突尖。头状花序球状，基部具四枚白色苞片。复合果成熟时红色，球状。花期5~7月，果期9~10月。

Deciduous trees, 3~10 m tall. Bark grayish brown, smooth. Leaf blade oppsite, abaxially often densely papillate and pubescent with appressed trichomes, axils of veins often with a cluster of white or brown soft long trichomes, veins 4 or 5 pairs, base abruptly acute to rounded, apex abruptly acuminate. Capitate cymes globose, with 4 white bracts on the base. Compound fruit red at maturity, globose. Fl. May - Jul., fr. Sep. - Oct..

树干　　Trunk
摄影：杨庆松　Photo by: Yang Qingsong

叶　　Leaves
摄影：杨庆松　Photo by: Yang Qingsong

果枝　　Fruiting branch
摄影：杨庆松　Photo by: Yang Qingsong

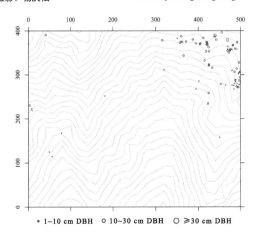

个体分布图 Distribution of individuals

径级分布表 DBH class

胸径等级 (Diameter class) (cm)	个体数 (No. of individuals in the plot)	比例 (Proportion) (%)
1~2	3	3.61
2~5	14	16.87
5~10	15	18.07
10~20	32	38.55
20~30	17	20.48
30~60	2	2.41
≥60	0	0.00

117 马银花   Mǎ yín huā | Azaleastra

Rhododendron ovatum (Lindl.) Planch. ex Maxim.
杜鹃花科 | Ericaceae

代码 (SpCode) = RHOOVA
个体数 (Individual number/20 hm^2) = 2735
最大胸径 (Max DBH) = 17.6 cm
重要值排序 (Importance value rank) = 15

常绿灌木，2~4(6) m高。叶柄长约8 mm，具狭翅，被短柔毛；叶片卵形至椭圆状卵形，3.5~5 × 1.9~2.5 cm；基部圆形或很少宽楔形；边缘稍弯曲；先端锐尖，具短尖头；正面无毛或短柔毛；侧脉及网脉不明显。花单生于枝顶叶腋，花冠紫白色，5裂，雄蕊5枚。蒴果宽卵形，密被灰褐色短柔毛。花期4~5月，果期7~10月。

Evergreen shrubs, 2~4(~6) m tall. Petiole ca. 8 mm, narrowly winged, pubescent; leaf blade ovate to oblong-elliptic, 3.5~5 × 1.9~2.5 cm; base rounded or rarely broadly cuneate; margin slightly curved; apex acute and mucronate; adaxial surface glabrous or pubescent; lateral and net veins inconspicuous abaxially. Flowers solitary at branch terminal leaf axil, corolla purple white, 5-lobed, stamens 5. Capsule broadly ovoid, densely gray-brown pubescent. Fl. Apr. - May, fr. Jul. - Oct..

树干   Trunk
摄影：杨庆松   Photo by: Yang Qingsong

枝叶   Branch and leaves
摄影：杨庆松   Photo by: Yang Qingsong

花枝   Flowering branch
摄影：杨庆松   Photo by: Yang Qingsong

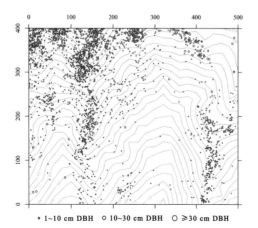

个体分布图   Distribution of individuals

径级分布表 DBH class

胸径区间 (Diameter class) (cm)	个体数 (No. of individuals in the plot)	比例 (Proportion) (%)
1~2	519	18.98
2~5	1460	53.38
5~10	682	24.94
10~20	74	2.71
20~30	0	0.00
30~60	0	0.00
≥60	0	0.00

118 杜鹃

Rhododendron simsii Planch.
杜鹃花科 | Ericaceae

代码 (SpCode) = RHOSIM
个体数 (Individual number/20 hm^2) = 156
最大胸径 (Max DBH) = 8.9 cm
重要值排序 (Importance value rank) = 71

落叶灌木，2 (5) m高；分枝多而纤细，密被亮棕褐色扁平糙伏毛。夏、冬叶不同。叶柄2~6 mm。叶常集生枝顶，卵形、椭圆状卵形或倒卵形至倒披针形；基部楔形或宽楔形；边缘稍外卷，具细齿；先端短渐尖。花2~3 (6) 朵簇生，花冠红或粉色，5裂。蒴果卵球形。花期4~5月，果期6~8月。

Deciduous shrubs, 2(~5) m tall; branches many and fine, densely shiny brown appressed-setose, setae flat. Summer and winter leaves different. Petiole 2~6 mm. Leaf blade set in the upper portion of the stem, ovate, elliptic-ovate or obovate to oblanceolate; base cuneate or broadly cuneate; margin slightly revolute, finely toothed; apex shortly acuminate. Inflorescence 2~3(~6) flowers tufted, corolla red or pink, 5-lobed. Capsule ovoid. Fl. Apr. - May, fr. Jun. - Aug..

枝叶　　Branches and leaves
摄影：杨庆松　　Photo by: Yang Qingsong

花　　Flowers
摄影：杨庆松　　Photo by: Yang Qingsong

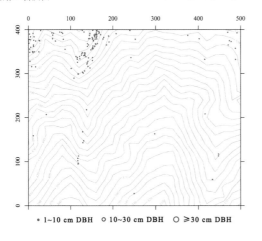

个体分布图　Distribution of individuals

径级分布表　DBH class

胸径等级 (Diameter class) (cm)	个体数 (No. of individuals in the plot)	比例 (Proportion) (%)
1~2	47	30.13
2~5	103	66.03
5~10	6	3.85
10~20	0	0.00
20~30	0	0.00
30~60	0	0.00
≥60	0	0.00

119 南烛 (乌饭树)

Nán zhú | Oriental Blueberry

Vaccinium bracteatum Thunb.
杜鹃花科 | Ericaceae

代码 (SpCode) = VACBRA

个体数 (Individual number/20 hm^2) = 70

最大胸径 (Max DBH) = 28.1 cm

重要值排序 (Importance value rank) = 77

常绿灌木或小乔木，2～6 (9) m 高，分枝多。叶散生；叶柄 2～8 mm，被微柔毛或无毛；叶片椭圆形、菱形或披针形，椭圆形，或披针形，很少倒卵形，薄革质，基部楔形、宽楔形或钝，边缘具小齿，先端锐尖，渐尖或长渐尖，很少圆形。总状花序顶生和腋生，花冠白色，筒状。浆果熟时紫黑色，被短柔毛。花期6～7月，果期8～10月。

Evergreen shrubs or small trees, 2~6(~9) m tall, much branched. Leaves scattered; petiole 2~8 mm, puberulous or glabrous; leaf blade elliptic, rhombic- or lanceolate-elliptic, or lanceolate, rarely obovate, thinly leathery, base cuneate, broadly cuneate, or obtuse, margin denticulate, apex acute, acuminate, rarely rounded or long acuminate. Inflorescences axillary and terminal, corolla white, rarely reddish, tubular or slightly urceolate, Berry dark purple in mature, pubescent. Fl. Jun. - Jul., fr. Aug. - Oct..

枝叶　Branch and leaves
摄影：杨庆松　Photo by: Yang Qingsong

花　Flowers
摄影：胡瑾瑾　Photo by: Hu Jinjin

果枝　Fruiting branches
摄影：杨庆松　Photo by: Yang Qingsong

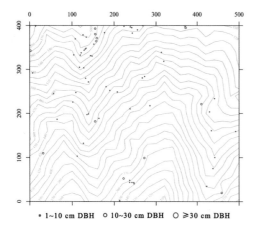

个体分布图　Distribution of individuals

径级分布表　DBH class

胸径区间 (Diameter class) (cm)	个体数 (No. of individuals in the plot)	比例 (Proportion) (%)
1～2	3	4.29
2～5	27	38.57
5～10	27	38.57
10～20	10	14.29
20～30	3	4.29
30～60	0	0.00
≥60	0	0.00

120 江南越桔

Jiāng nán yuè jú | China Blueberry

Vaccinium mandarinorum Diels
杜鹃花科 | Ericaceae

代码 (SpCode) = VACMAN
个体数 (Individual number/20 hm²) = 167
最大胸径 (Max DBH) = 11.8 cm
重要值排序 (Importance value rank) = 67

常绿灌木或小乔木，1～4 (7) m高。小枝圆柱状，无毛，嫩枝有时密被微柔毛。叶散生；叶柄3～8 mm，无毛或被短柔毛；叶片革质或薄革质，两面通常无毛，基部楔形或圆形，边缘具牙齿，先端锐突尖约1.5 cm。总状花序腋生兼顶生；花冠白色，管状，5浅裂。浆果熟时紫黑色，无毛。花期4～6月，果期6～10月。

Evergreen shrubs or small trees, 1~4(~7) m tall. Twigs terete, glabrous, young branch sometimes densely puberulous. Leaves scattered; petiole 3~8 mm, glabrous or pubescent; leaf blade leathery to thinly leathery, both surfaces usually glabrous, base cuneate to rounded, margin dentate, apex acute to abruptly acuminate for ca. 1.5 cm. Inflorescence raceme, axillary and terminal; corolla white, bube-like, 5-lobed. Berry dark purple in mature, glabrous. Fl. Apr. - Jun., fr. Jun. - Oct..

枝叶 Branch and leaves
摄影：杨庆松 Photo by: Yang Qingsong

花枝 Flowering branch
摄影：汪远 Photo by: Wang Yuan

果枝 Fruiting branch
摄影：汪远 Photo by: Wang Yuan

径级分布表 DBH class

胸径等级 (Diameter class) (cm)	个体数 (No. of individuals in the plot)	比例 (Proportion) (%)
1～2	33	19.76
2～5	78	46.71
5～10	55	32.93
10～20	1	0.60
20～30	0	0.00
30～60	0	0.00
≥60	0	0.00

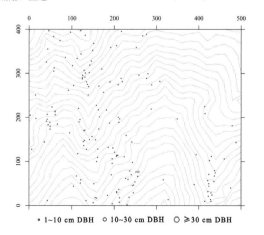

个体分布图 Distribution of individuals

121 刺毛越桔

Cì máo yuè jú | Spinehairy Blueberry

Vaccinium trichocladum Merr. et Metc.
杜鹃花科 | Ericaceae

代码 (SpCode) = VACTRI
个体数 (Individual number/20 hm^2) = 75
最大胸径 (Max DBH) = 9.2 cm
重要值排序 (Importance value rank) = 89

常绿灌木，偶见小乔木，高3~8 m。小枝圆形，密被或疏生具腺具刚毛和短硬毛，后脱落。叶散生；叶柄2~4 mm，毛被同枝；叶片卵状披针形，4~9 × 2~3 cm，薄革质，基部圆形或微呈心形，边缘具牙齿或近全缘，先端渐尖。浆果红色，球状，5~6 mm直径。花期4月，果期5~9月。

Evergreen shrubs, occasionally small trees, 3~8 m tall. Twigs rounded, densely or sparsely glandular setose and hispidulous, glabrescent. Leaves scattered; petiole 2~4 mm, indumentum same as twigs; leaf blade ovate-lanceolate, 4~9 × 2~3 cm, thinly leathery, base rounded or slightly cordate, margin plane, spiniform-dentate or subentire, apex acuminate. Berry red, globose, 5~6 mm in diam. Fl. Apr., fr. May - Sep..

茎干　Stem
摄影：杨庆松　Photo by: Yang Qingsong

花枝　Flowering branches
摄影：杨庆松　Photo by: Yang Qingsong

枝叶　Branch and leaves
摄影：杨庆松　Photo by: Yang Qingsong

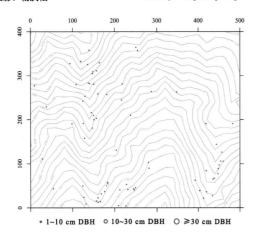
个体分布图　Distribution of individuals

径级分布表 DBH class

胸径区间 (Diameter class) (cm)	个体数 (No. of individuals in the plot)	比例 (Proportion) (%)
1~2	21	28.00
2~5	43	57.33
5~10	11	14.67
10~20	0	0.00
20~30	0	0.00
30~60	0	0.00
≥60	0	0.00

122 山柿

Shān shì | Mountain Persimmon

Diospyros japonica Siebold et Zucc.
柿科 | Ebenaceae

代码 (SpCode) = DIOJAP
个体数 (Individual number/20 hm²) = 88
最大胸径 (Max DBH) = 51.2 cm
重要值排序 (Importance value rank) = 38

落叶乔木，高达17 m。小枝暗褐色到黑褐色，无毛。叶柄1.2~2.5 cm；叶片椭圆形至披针形，7.5~17.5×3.5~7.5 cm，薄革质，无毛或背面疏生贴伏短柔毛，基部圆形至截形，先端渐尖，侧脉7或8每侧，网状细脉密集。雄花小，生于聚伞花序上，雌花单生，花萼绿色，花冠淡黄色。浆果橙黄色，后变为红色，球状到扁球形，1.5~2 (3) cm直径，有白霜。花期4~7月，果期9~11月。

Deciduous trees, to 17 m tall. Branchlets dark brown to blackish brown, glabrous. Petiole 1.2~2.5 cm; leaf blade elliptic to lanceolate, 7.5~17.5 × 3.5~7.5 cm, thinly leathery, glabrous or sparsely appressed pubescent, abaxially glaucous, base rounded to truncate, apex acuminate, lateral veins 7 or 8 per side, reticulate veinlets dense. Staminate flowers in cymes, Pistillate flowers solitary, corolla yellowish Berries orange-yellow, becoming red, globose to depressed globose, 1.5~2(~3) cm in diam., glaucous. Fl. Apr. - Jul., fr. Sep. - Nov..

树干 Trunk
摄影：杨庆松 Photo by: Yang Qingsong

叶 Leaf
摄影：杨庆松 Photo by: Yang Qingsong

果枝 Fruiting branch
摄影：杨庆松 Photo by: Yang Qingsong

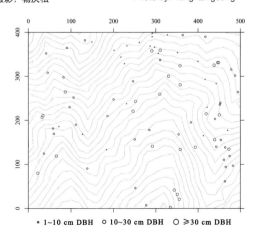

个体分布图 Distribution of individuals

径级分布表 DBH class

胸径等级 (Diameter class) (cm)	个体数 (No. of individuals in the plot)	比例 (Proportion) (%)
1~2	8	9.09
2~5	4	4.55
5~10	11	12.50
10~20	21	23.86
20~30	17	19.32
30~60	27	30.68
≥60	0	0.00

123 油柿   Yóu shì | Oily Persimmon

Diospyros oleifera Cheng
柿科 | Ebenaceae

代码 (SpCode) = DIOOLE
个体数 (Individual number/20 hm^2) = 40
最大胸径 (Max DBH) = 65.6 cm
重要值排序 (Importance value rank) = 69

落叶乔木，14 m高。树皮灰黑色到灰褐色，呈大的薄片状剥落，露出下面的白色树皮。分枝疏生柔毛到后脱落。叶柄6～10 mm；叶片长圆形至长圆状披针形，很少倒卵形，6.5～17×3.5～10 cm，纸质，基部圆形至近圆形，稍偏斜，侧脉每边7～9。浆果暗黄色，卵形至扁球形。花期4～5月，果期8～10月。

Trees, to 14 m tall, deciduous. Bark dark gray to grayish brown, peeling in large thin flakes to reveal white bark below. Branches sparsely villose to glabrescent. Petiole 6~10 mm; leaf blade oblong to oblong-lanceolate, rarely obovate, 6.5~17 × 3.5~10 cm, papery, base rounded to subrounded and slightly oblique, lateral veins 7~9 per side. Berries dark yellow, ovoid to depressed globose. Fl. Apr. - May, fr. Aug. - Oct..

树干   Trunk
摄影：杨庆松   Photo by: Yang Qingsong

叶   Leaves
摄影：杨庆松   Photo by: Yang Qingsong

果   Fruit
摄影：杨庆松   Photo by: Yang Qingsong

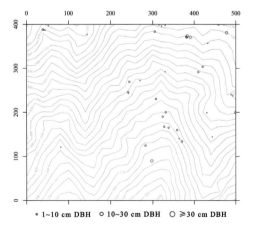
个体分布图 Distribution of individuals

径级分布表 DBH class

胸径区间 (Diameter class) (cm)	个体数 (No. of individuals in the plot)	比例 (Proportion) (%)
1～2	6	15.00
2～5	4	10.00
5～10	8	20.00
10～20	9	22.50
20～30	8	20.00
30～60	4	10.00
≥60	1	2.50

124 薄叶山矾

Báo yè shān fán | Thinleaf Sweetleaf

Symplocos anomala Brand
山矾科 | Symplocaceae

代码 (SpCode) = SYMANO
个体数 (Individual number/20 hm^2) = 3273
最大胸径 (Max DBH) = 12.1 cm
重要值排序 (Importance value rank) = 13

常绿灌木或小乔木。叶柄2～7 mm；叶薄革质，狭椭圆形、椭圆形或卵形，背面无毛或很少嫩叶细伏毛，基部渐狭楔形，全缘或细腺齿，先端渐尖，中脉和侧脉正面突出，侧脉5～11对。总状花序腋生，花冠白色。核果棕色，长圆球形。花果期4～12月。

Evergreen shrubs or small trees. Petiole 2~7 mm; leaf blade thinly leathery, narrowly elliptic, elliptic or ovate, base attenuate-cuneate, margin entire or finely glandular dentate, apex acuminate, midvein and lateral veins adaxially prominent, lateral veins 5~11 pairs. Inflorescence racemes, axillary, corolla white. Drupes brown, oblong-globose. Fl. and fr. Apr. - Dec..

叶　　　　Leaves
摄影：杨庆松　Photo by: Yang Qingsong

枝叶　　　　Branch and leaves
摄影：杨庆松　Photo by: Yang Qingsong

花枝　　　　Flowering branch
摄影：杨庆松　Photo by: Yang Qingsong

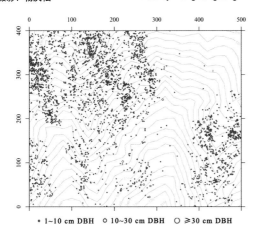

个体分布图　Distribution of individuals

径级分布表　DBH class

胸径等级 (Diameter class) (cm)	个体数 (No. of individuals in the plot)	比例 (Proportion) (%)
1～2	1071	32.72
2～5	1760	53.77
5～10	420	12.83
10～20	22	0.67
20～30	0	0.00
30～60	0	0.00
≥60	0	0.00

125 黄牛奶树

Huáng niú nǎi shù | Laurel Sweetleaf

Symplocos cochinchinensis var. *laurina* (Retz.) Noot.
山矾科 | Symplocaceae

代码 (SpCode) = SYMCOC
个体数 (Individual number/20 hm^2) = 1022
最大胸径 (Max DBH) = 41.9 cm
重要值排序 (Importance value rank) = 26

常绿小乔木。小枝无毛；芽棕色短柔毛。叶柄 1~1.5 cm；叶片卵状椭圆形至倒卵状椭圆形，6~21 × 2~8 cm，革质，无毛，叶边缘有细锯齿，先端渐尖或急尖。穗状花序 3~14 cm。核果球状，4~6 mm 直径。花期 8~12 月，果期翌年 3~6 月。

Evergreen small trees. Branchlets glabrous; buds brown pubescent. Petiole 1~1.5 cm; leaf blade ovate-elliptic to obovate-elliptic, 6~21 × 2~8 cm, leathery, glabrous, margin serrulate, apex acuminate to acute; leaf blade lateral veins 5~9 pairs, not strictly parallel, anastomosing at some distance from margin. Spikes 3~14 cm. Drupes globose, 4~6 mm in diam. Fl. Aug. - Dec., fr. Mar. - Jun. of following year.

树干　Trunk
摄影：杨庆松　Photo by: Yang Qingsong

枝叶　Branch and leaves
摄影：雷霄　Photo by: Lei Xiao

芽　Sprout
摄影：杨庆松　Photo by: Yang Qingsong

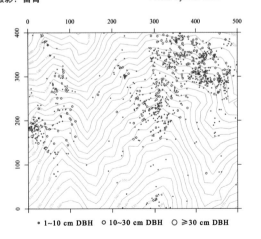

个体分布图　Distribution of individuals

径级分布表　DBH class

胸径区间 (Diameter class) (cm)	个体数 (No. of individuals in the plot)	比例 (Proportion) (%)
1~2	199	19.47
2~5	336	32.88
5~10	254	24.85
10~20	209	20.45
20~30	23	2.25
30~60	1	0.10
≥60	0	0.00

126 黑山山矾 (海桐山矾)   Hēi shān shān fán | Heishan Sweetleaf

Symplocos heishanensis Hayata
山矾科 | Symplocaceae

代码 (SpCode) = SYMHEI
个体数 (Individual number/20 hm^2) = 89
最大胸径 (Max DBH) = 22.0 cm
重要值排序 (Importance value rank) = 63

常绿灌木或乔木，高3～15 m。幼枝深棕色，无毛或被微柔毛；老小枝黑色；芽稍短柔毛。叶柄0.5～1.5 cm；叶片狭椭圆形至倒卵状椭圆形，6～12×1.1～4 cm，革质，无毛，基部楔形，边缘全缘或波状齿，先端尾状渐尖，侧脉9～14对。核果暗紫色，长壶状。花期2～5月，果期6～9月。

Evergreen shurbs or trees, to 3~15 m tall. Young branchlets dark brown, glabrous or puberulent; old branchlets black; buds minutely pubescent. Petiole 0.5~1.5 cm; leaf blade narrowly elliptic to obovate-elliptic, 6~12 × 1.1~4 cm, leathery, glabrous, base cuneate, margin entire or sinuous-dentate, apex caudate-acuminate, lateral veins 9~14 pairs. Drupes dark purplish, long ampulliform. Fl. Feb. - May., fr. Jun. - Sep..

叶   Leaves
摄影：杨庆松   Photo by: Yang Qingsong

枝叶   Branch and leaves
摄影：杨庆松   Photo by: Yang Qingsong

花枝   Flowering branches
摄影：杨庆松   Photo by: Yang Qingsong

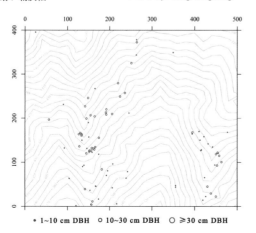
个体分布图 Distribution of individuals

径级分布表 DBH class

胸径等级 (Diameter class) (cm)	个体数 (No. of individuals in the plot)	比例 (Proportion) (%)
1～2	8	8.99
2～5	7	7.87
5～10	31	34.83
10～20	41	46.07
20～30	2	2.25
30～60	0	0.00
≥60	0	0.00

127 光叶山矾 (披针叶山矾)   Guāng yè shān fán | Smoothleaf Sweetleaf

Symplocos lancifolia Sieb. et Zucc.
山矾科 | Symplocaceae

代码 (SpCode) = SYMLAN
个体数 (Individual number/20 hm^2) = 354
最大胸径 (Max DBH) = 25.2 cm
重要值排序 (Importance value rank) = 41

常绿灌木或乔木，高可达20 m。芽、嫩枝、嫩叶背面脉上、花序均被黄褐色柔毛。叶柄1～5 mm；叶片近膜质或纸质，基部渐狭楔形，边缘具细圆齿，先端尾状渐尖。穗状花序腋生，长1～4 cm，花冠淡黄色。核果椭球型至近球形，3-5×2～5 mm。花期3～11月，果期6～12月。

Evergreen shrubs or trees, to 20 m. Buds, young branchlets, and inflorescence axes appressed to patently hairy. Petiole 1~5 mm; leaf blade submembranous to papery, base attenuate-cuneate, margin finely crenate to dentate, apex caudate-acuminate. Inflorescence spike, 1~4 cm, corolla pale yellow. Drupes ellipsoid to subglobose, 3~5 × 2~5 mm. Fl. Mar. - Nov., fr. Jun. - Dec..

树干 Trunk
摄影：杨庆松 Photo by: Yang Qingsong

枝叶 Branches and leaves
摄影：雷霄 Photo by: Lei Xiao

果枝 Fruiting branches
摄影：杨庆松 Photo by: Yang Qingsong

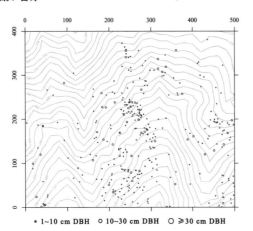

个体分布图 Distribution of individuals

径级分布表 DBH class

胸径区间 (Diameter class) (cm)	个体数 (No. of individuals in the plot)	比例 (Proportion) (%)
1～2	76	21.47
2～5	106	29.94
5～10	122	34.46
10～20	49	13.84
20～30	1	0.28
30～60	0	0.00
≥60	0	0.00

128 光亮山矾 (四川山矾)　　　　　Guāng liàng shān fǎn | Sichuan Sweetleaf

Symplocos lucida (Thunb.) Siebold et Zucc.
山矾科 | Symplocaceae

代码 (SpCode) = SYMLUC
个体数 (Individual number/20 hm^2) = 1670
最大胸径 (Max DBH) = 24.8 cm
重要值排序 (Importance value rank) = 22

常绿灌木或乔木。小枝黄绿色，具棱，无毛。叶柄0.5～1.5 cm；叶片长圆形至狭椭圆形，5～13×2～5 cm，革质，无毛，基部楔形，边缘反卷，具尖锯齿，先端长渐尖或急尖，正面中脉突出。总状花序或穗状花序短缩成密伞花序，有花多朵，生于叶腋，花冠白色。核果卵状椭圆形。花期3～4月，果期5～6月。

Evergreen shrubs or trees. Branchlets mostly yellowish green, slightly angled to winged, glabrous. Petiole 0.5~1.5 cm; leaf blade oblong to narrowly elliptic, 5~13 × 2~5 cm, leathery, glabrous, base cuneate, margin revolute and sharply dentate, apex long acuminate to acute, midvein adaxially prominent. Inflorescence spike or raceme, shortened to a fascicle, with many flowers, axillary, corolla white. Drupes ovoid ellipsoid. Fl. Mar. - Apr., fr. May - Jun..

树皮　Bark
摄影：杨庆松　Photo by: Yang Qingsong

枝叶　Branches and leaves
摄影：杨庆松　Photo by: Yang Qingsong

花枝　Flowering branch
摄影：杨庆松　Photo by: Yang Qingsong

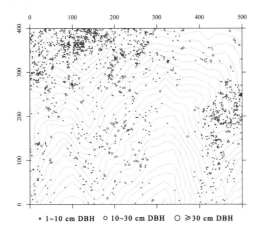
个体分布图　Distribution of individuals

径级分布表　DBH class

胸径等级 (Diameter class) (cm)	个体数 (No. of individuals in the plot)	比例 (Proportion) (%)
1～2	332	19.88
2～5	655	39.22
5～10	539	32.28
10～20	143	8.56
20～30	1	0.06
30～60	0	0.00
≥60	0	0.00

129 白檀

Bái tán | Sapphireberry Sweetleaf

Symplocos paniculata (Thunb.) Miq.
山矾科 | Symplocaceae

代码 (SpCode) = SYMPAN
个体数 (Individual number/20 hm^2) = 1
最大胸径 (Max DBH) = 12.0 cm
重要值排序 (Importance value rank) = 138

落叶灌木或小乔木。叶膜质或薄纸质，阔倒卵形、椭圆状倒卵形或卵形，先端急尖或渐尖，基部阔楔形或近圆形，边缘有细尖锯齿。圆锥花序顶生，花冠白色。核果熟时蓝色，卵状球形。花期4~6月，果期9~11月。

Shrubs or small trees, deciduous. Leaf blade ovate, elliptic-obovate, or broadly obovate, membranous to thinly papery, base broadly cuneate to subcordate, margin sharply glandular dentate, apex acuminate to acute. Panicles terminal, corolla white. Drupes bluish, ovate-globose. Fl. Apr. - Jun., fr. Sep. - Nov..

枝叶　Branch and leaves
摄影：杨庆松　Photo by: Yang Qingsong

果枝　Fruiting branches
摄影：雷霄　Photo by: Lei Xiao

果　Fruits
摄影：杨庆松　Photo by: Yang Qingsong

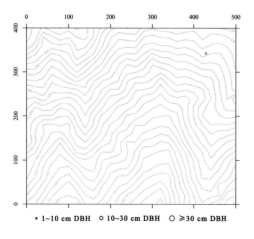

个体分布图 Distribution of individuals

径级分布表 DBH class

胸径区间 (Diameter class) (cm)	个体数 (No. of individuals in the plot)	比例 (Proportion) (%)
1~2	0	0.00
2~5	0	0.00
5~10	0	0.00
10~20	1	100.00
20~30	0	0.00
30~60	0	0.00
≥60	0	0.00

130 老鼠矢

Lǎo shǔ shǐ | Starshape Sweetleaf

Symplocos stellaris Brand
山矾科 | Symplocaceae

代码 (SpCode) = SYMSTE
个体数 (Individual number/20 hm^2) = 232
最大胸径 (Max DBH) = 9.7 cm
重要值排序 (Importance value rank) = 58

常绿灌木或小乔木，小枝粗；芽和嫩枝被红褐色绒毛。叶厚革质，叶面有光泽，叶背粉褐色，披针状椭圆形或狭长圆状椭圆形，通常全缘。团伞花序；苞片和小苞片宿存，密被绒毛。花冠白色，长5～8 mm；核果狭卵状圆柱形，长约1 cm。花期4～5月，果期6～9月。

Evergreen shrubs or small trees; branchlets strong. Buds and young branchlets reddish brown tomentellous. Leaf blade narrowly oblong-elliptic to narrowly obovate, thickly leathery, glabrous, abaxially smooth, often light colored, margin usually entire. Inflorescences a glomerule or condensed spike; bracts and bractlets persistent, densely tomentose; corolla white, 5~8 mm. Drupes narrowly ovoid-cylindric, ca. 1 cm. Fl. Apr. - May, fr. Jun. - Sep..

花枝　Flowering branch
摄影：杨庆松　Photo by: Yang Qingsong

枝叶　Branch and leaves
摄影：杨庆松　Photo by: Yang Qingsong

花　Flowers
摄影：杨庆松　Photo by: Yang Qingsong

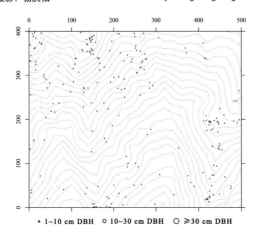
个体分布图　Distribution of individuals

径级分布表 DBH class

胸径等级 (Diameter class) (cm)	个体数 (No. of individuals in the plot)	比例 (Proportion) (%)
1～2	101	43.53
2～5	85	36.64
5～10	46	19.83
10～20	0	0.00
20～30	0	0.00
30～60	0	0.00
≥60	0	0.00

131 山矾 — Shān fán | Sweetleaf

Symplocos sumuntia Buch.-Ham. ex D. Don.
山矾科 | Symplocaceae

代码 (SpCode) = SYMSUM
个体数 (Individual number/20 hm^2) = 1480
最大胸径 (Max DBH) = 13.8 cm
重要值排序 (Importance value rank) = 28

常绿灌木，嫩枝褐色。叶薄革质，卵形、狭倒卵形、倒披针状椭圆形，两面无毛，有时背面有毛，先端常呈尾状渐尖，基部楔形或圆形，边缘具浅锯齿或波状齿，有时近全缘。总状花序，花冠白色。核果卵状坛形，顶端宿萼裂片直立。花期2～4月，果期6～11月。

Evergreen shrubs, young branchlets brown. Leaf blade elliptic, narrowly ovate, or ovate, thinly leathery, both surfaces glabrous, sometimes abaxially hairy, base cuneate to rounded, margin slightly serrate, sinuolate-dentate, or rarely subentire, apex caudate. Racemes. Corolla white. Drupes ampulliform to ovoid, apex with persistent erect calyx lobes. Fl. Feb. - Apr., fr. Jun. - Nov..

花　Flowers
摄影：杨庆松　Photo by: Yang Qingsong

枝叶　Branches and leaves
摄影：杨庆松　Photo by: Yang Qingsong

果枝　Fruiting branch
摄影：杨庆松　Photo by: Yang Qingsong

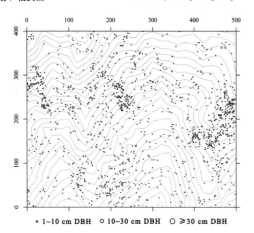
个体分布图　Distribution of individuals

径级分布表　DBH class

胸径区间 (Diameter class) (cm)	个体数 (No. of individuals in the plot)	比例 (Proportion) (%)
1～2	586	39.59
2～5	767	51.82
5～10	120	8.11
10～20	7	0.47
20～30	0	0.00
30～60	0	0.00
≥60	0	0.00

132 赤杨叶 (拟赤杨)

Chì yáng yè | Fortune Chinabells

Alniphyllum fortunei (Hemsl.) Makino
安息香科 | Styracaceae

代码 (SpCode) = ALNFOR
个体数 (Individual number/20 hm^2) = 623
最大胸径 (Max DBH) = 36.8 cm
重要值排序 (Importance value rank) = 30

落叶乔木，高达20 m，树干通直，树皮灰褐色，有不规则细纵皱纹。小枝带褐色星状短柔毛。叶纸质，顶端急尖至渐尖，基部宽楔形或楔形，边缘具疏离硬质锯齿，两面疏生至密被褐色星状短柔毛或星状绒毛，有时脱落变为无毛。总状花序或圆锥花序，顶生或腋生。果实长圆形或长椭圆形；种子多数，长5～7 mm。花期4～7月，果期8～10月。

Deciduous trees to 20 m tall. Trunk straight with irregular longitudinal crack; bark gray-brown. Branchlets brownish stellate pubescent. Leaf blade, papery, sparsely to densely brown stellate pubescent to tomentose, abaxially sometimes glabrescent, base broadly cuneate to cuneate, margin remotely serrate, apex acute to acuminate. Inflorescences terminal or axillary, racemes or panicles. Fruit oblong to ellipsoid. Seeds many, 5~7 mm. Fl. Apr. - Jul., fr. Aug. - Oct..

树干　　Trunk
摄影：杨庆松　Photo by: Yang Qingsong

果枝　　Fruiting branch
摄影：严靖　Photo by: Yan Jing

叶　　Leaves
摄影：杨庆松　Photo by: Yang Qingsong

径级分布表 DBH class

胸径等级 (Diameter class) (cm)	个体数 (No. of individuals in the plot)	比例 (Proportion) (%)
1～2	141	22.63
2～5	131	21.03
5～10	166	26.65
10～20	157	25.20
20～30	21	3.37
30～60	7	1.12
≥60	0	0.00

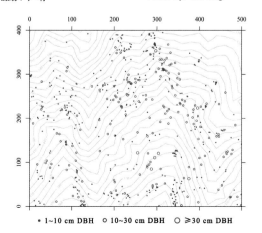

个体分布图　Distribution of individuals

133 赛山梅

Sài shān méi | Muddy Snowbell

Styrax confusus Hemsl.
安息香科 | Styracaceae

代码 (SpCode) = STYCON
个体数 (Individual number/20 hm^2) = 491
最大胸径 (Max DBH) = 19.1 cm
重要值排序 (Importance value rank) = 40

落叶小乔木，高2~8 m。小枝密被黄褐色星状短柔毛，圆柱形，紫红色。叶革质或近革质，互生，椭圆形、长圆状椭圆形或倒卵状椭圆形，边缘有细锯齿。总状花序顶生，有花3~8朵，花白色。果实近球形或倒卵形，外面密被灰黄色星状绒毛和星状长柔毛；种子倒卵形，褐色，平滑或具深皱纹。花期4~6月，果期9~11月。

Deciduous small trees 2~8 m tall. Branchlets densely brownish stellate pubescent, cylinder, purplish. Leaves alternate; leaf blade narrowly oblong, obovate-elliptic, or oblong-elliptic, leathery to almost leathery, margin serrulate. Racemes terminal, 3~8-flowered, corolla white. Fruit subglobose to obovoid, densely yellowish stellate tomentose. Seeds brown, obovoid, smooth or deeply rugose. Fl. Apr. - Jun., fr. Sep. - Nov..

枝叶　Branch and leaves
摄影：杨庆松　Photo by: Yang Qingsong

花枝　Flowering branches
摄影：杨庆松　Photo by: Yang Qingsong

果枝　Fruiting branches
摄影：杨庆松　Photo by: Yang Qingsong

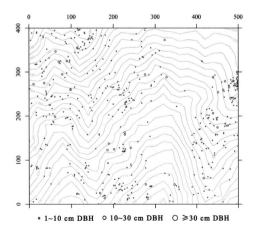
个体分布图　Distribution of individuals

径级分布表　DBH class

胸径区间 (Diameter class) (cm)	个体数 (No. of individuals in the plot)	比例 (Proportion) (%)
1~2	117	23.83
2~5	219	44.60
5~10	118	24.03
10~20	37	7.54
20~30	0	0.00
30~60	0	0.00
≥60	0	0.00

134 栓叶安息香 (红皮树)

Shuān yè ān xī xiāng | Corkleaf Snowbell

Styrax suberifolius Hook. et Arn.
安息香科 | Styracaceae

代码 (SpCode) = STYSUB
个体数 (Individual number/20 hm^2) = 5
最大胸径 (Max DBH) = 18.6 cm
重要值排序 (Importance value rank) = 118

常绿乔木，高4~20 m，树皮红褐色；嫩枝被锈褐色星状绒毛。叶互生，革质，椭圆形、长椭圆形或椭圆状披针形，顶端渐尖，尖头有时稍弯，基部楔形，边近全缘，上面无毛或仅中脉疏被星状毛，下面密被黄褐色至灰褐色星状绒毛；叶柄长1~1.5 (2) cm。总状花序或圆锥花序，顶生或腋生，花多数；种子褐色，无毛。花期3~5月，果期8~11月。

Evergreen trees 4~20 m tall. Bark reddish brown. Branchlets stellate tomentose. Leaves alternate; petiole 1~1.5(~2) cm, nearly 4-angled; leaf blade elliptic, oblong, or elliptic-lanceolate, leathery, abaxially densely brownish stellate tomentose, adaxially subglabrous or primary vein sparsely stellate pubescent, base cuneate, margin subentire, apex acuminate and sometimes slightly curved. Inflorescences terminal or axillary, racemes or panicles, many-flowered, Seeds brown, glabrous. Fl. Mar. - May, fr. Aug. - Nov..

树干　Trunk
摄影：杨庆松　Photo by: Yang Qingsong

枝叶　Branch and leaves
摄影：杨庆松　Photo by: Yang Qingsong

叶背　Leaf abaxial surface
摄影：葛斌杰　Photo by: Ge Binjie

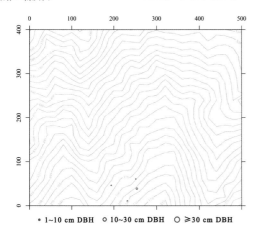

个体分布图　Distribution of individuals

径级分布表 DBH class

胸径等级 (Diameter class) (cm)	个体数 (No. of individuals in the plot)	比例 (Proportion) (%)
1~2	1	20.00
2~5	0	0.00
5~10	3	60.00
10~20	1	20.00
20~30	0	0.00
30~60	0	0.00
≥60	0	0.00

135 苦枥木

Kǔ lì mù | Insular Ash

Fraxinus insularis Hemsl.
木犀科 | Oleaceae

代码 (SpCode) = FRAINS
个体数 (Individual number/20 hm^2) = 412
最大胸径 (Max DBH) = 45.5 cm
重要值排序 (Importance value rank) = 33

落叶大乔木，高20～30 m；树皮灰色，平滑。嫩枝扁平，节膨大。羽状复叶长10～30 cm；小叶 (3) 5～7枚，嫩时纸质，后期变硬纸质或革质，先端急尖、渐尖以至尾尖，基部楔形至钝圆，叶缘具浅锯齿，或中部以下近全缘，两面无毛。圆锥花序生于当年生枝端，顶生及侧生叶腋，多花，叶后开放；花冠白色。翅果红色至褐色，长匙形。花期4～5月，果期7～9月。

Deciduous trees 20~30 m. Bark gray, smooth. Branchlets compressed when young. Leaves 10~30 cm; leaflets 3~5(~7), papery becoming lathery, glabrous, base cuneate or blunt, serrate or entire below the middle, apex acute, acuminate, to caudate. Panicles terminal or terminal and lateral, 20~30 cm, many flowered. Flowers appearing after leaves. Corolla white. Samara red to brown, long spatulate. Fl. Apr. - May, fr. Jul. - Sep..

枝叶　Branch and leaves
摄影：杨庆松　Photo by: Yang Qingsong

果枝　Fruiting branch
摄影：杨庆松　Photo by: Yang Qingsong

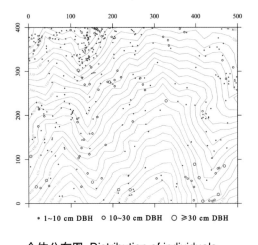

个体分布图 Distribution of individuals

径级分布表 DBH class

胸径区间 (Diameter class) (cm)	个体数 (No. of individuals in the plot)	比例 (Proportion) (%)
1～2	95	23.06
2～5	166	40.29
5～10	70	16.99
10～20	37	8.98
20～30	30	7.28
30～60	14	3.40
≥60	0	0.00

136 小蜡 — Xiǎo là | Small Privet

Ligustrum sinense Lour.
木犀科 | Oleaceae

代码 (SpCode) = LIGSIN
个体数 (Individual number/20 hm²) = 11
最大胸径 (Max DBH) = 6.0 cm
重要值排序 (Importance value rank) = 116

落叶灌木或小乔木，高 2～4(7) m。小枝圆柱形。叶片纸质或薄革质，卵形、椭圆状卵形、长圆形、长圆状椭圆形至披针形，或近圆形，先端锐尖、短渐尖至渐尖，或钝而微凹，基部宽楔形至近圆形，或为楔形；叶柄长 2～8 mm。圆锥花序顶生或腋生，塔形。果近球形，径 5～8 mm。花期 3～6 月，果期 9～12 月。

Shrubs or small trees 2~4(~7) m, deciduous. Branchlets terete. Petiole 2~8 mm; leaf blade ovate, oblong, elliptic to lanceolate, or suborbicular, papery to somewhat leathery, base cuneate to subrounded, apex acute to acuminate, sometimes obtuse and retuse. Panicles terminal or axillary. Fruit subglobose, 5~8 mm in diam. Fl. Mar. - Jun., fr. Sep. - Dec..

茎干　Stem
摄影：杨庆松　Photo by: Yang Qingsong

枝叶　Branch and leaves
摄影：杨庆松　Photo by: Yang Qingsong

果枝　Fruiting branch
摄影：杨庆松　Photo by: Yang Qingsong

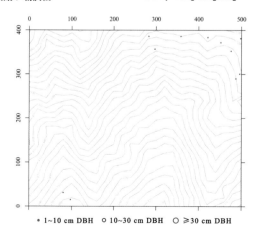
个体分布图　Distribution of individuals

径级分布表 DBH class

胸径等级 (Diameter class) (cm)	个体数 (No. of individuals in the plot)	比例 (Proportion) (%)
1～2	8	72.73
2～5	2	18.18
5～10	1	9.09
10～20	0	0.00
20～30	0	0.00
30～60	0	0.00
≥60	0	0.00

137 宁波木犀 (华东木犀)

Níng bō mù xī | Cooper Osmanther

Osmanthus cooperi Hemsl.
木犀科 | Oleaceae

代码 (SpCode) = OSMCOO
个体数 (Individual number/20 hm^2) = 503
最大胸径 (Max DBH) = 45.2 cm
重要值排序 (Importance value rank) = 32

常绿乔木或小乔木，高10 m。小枝灰白色，幼枝黄白色，具较多皮孔。叶片革质，椭圆形或倒卵形，先端渐尖，稍呈尾状，基部宽楔形至圆形，全缘，中脉在上面凹入，在下面凸起；叶柄长1～2 cm。花序簇生于叶腋，每腋内有花4～12朵；花冠白色。核果长1.5～2 cm，呈蓝黑色。花期9～10月，果期翌年5～6月。

Evergreen trees or small trees, 10 m tall. Petiole 1~2 cm; leaf blade elliptic or obovate, leathery, base broadly cuneate to rounded, margin entire, apex acuminate and slightly caudate; impressed adaxially, raised abaxially. Cymes fascicled in leaf axils, 4~12-flowered. Corolla white. Drupe blue-black, 1.5~2 cm. Fl. Sep. - Oct., fr. May - Jun. of following year.

树干　Trunk
摄影：杨庆松　Photo by: Yang Qingsong

枝叶　Branch and leaves
摄影：杨庆松　Photo by: Yang Qingsong

花枝　Flowering branch
摄影：杨庆松　Photo by: Yang Qingsong

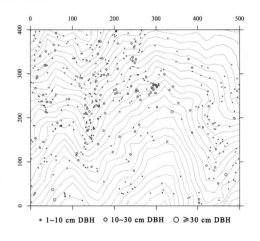

个体分布图　Distribution of individuals

径级分布表 DBH class

胸径区间 (Diameter class) (cm)	个体数 (No. of individuals in the plot)	比例 (Proportion) (%)
1～2	103	20.48
2～5	162	32.21
5～10	122	24.25
10～20	91	18.09
20～30	21	4.17
30～60	4	0.80
≥60	0	0.00

138 木犀 — Mù xī | Sweet Osmanther

Osmanthus fragrans (Thunb.) Lour.
木犀科 | Oleaceae

代码 (SpCode) = OSMFRA
个体数 (Individual number/20 hm^2) = 44
最大胸径 (Max DBH) = 16.4 cm
重要值排序 (Importance value rank) = 95

常绿乔木或灌木，高3～5(18) m。叶片革质，椭圆形、长椭圆形或椭圆状披针形，全缘或通常上半部具细锯齿，两面无毛，中脉在上面凹入，下面凸起。聚伞花序簇生于叶腋，有花多朵,花极芳香；花冠黄白色、淡黄色、黄色或桔红色。果歪斜，椭圆形，长1～1.5 cm，呈紫黑色。花期9～10月上旬，果期翌年3月。

Evergreen trees or shrubs 3~5(~18) m, glabrous. Leaf blade elliptic to elliptic-lanceolate, margin entire or usually serrulate along distal half; adaxially impressed and abaxially raised. Cymes fascicled in leaf axils, many flowered, very fragrant. Corolla yellowish, yellow, or orange. Drupe purple-black, ellipsoid, oblique, 1~1.5 cm. Fl. Sep. - early Oct., fr. Mar. of following year.

果枝　　Fruiting branch
摄影：王樟华　　Photo by: Wang Zhanghua

枝叶　　Branch and leaves
摄影：杨庆松　　Photo by: Yang Qingsong

花枝　　Flowering branch
摄影：杨庆松　　Photo by: Yang Qingsong

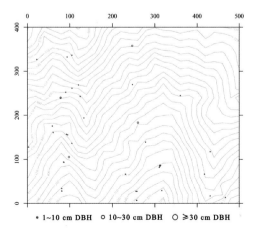
个体分布图　Distribution of individuals

径级分布表　DBH class

胸径等级 (Diameter class) (cm)	个体数 (No. of individuals in the plot)	比例 (Proportion) (%)
1～2	13	29.55
2～5	20	45.45
5～10	7	15.91
10～20	4	9.09
20～30	0	0.00
30～60	0	0.00
≥60	0	0.00

139 醉鱼草

Zuì yú cǎo | Lindley Summerlilic

Buddleja lindleyana Fortune
马钱科 | Loganiaceae

代码 (SpCode) = BUDLIN
个体数 (Individual number/20 hm^2) = 3
最大胸径 (Max DBH) = 1.6 cm
重要值排序 (Importance value rank) = 135

落叶灌木，高1～3 m。茎皮褐色；小枝具四棱，棱上略有窄翅；幼枝、叶片下面、叶柄、花序、苞片及小苞片均密被星状短绒毛和腺毛。叶片膜质、卵形、椭圆形至长圆状披针形，顶端渐尖，边缘全缘或具有波状齿。穗状聚伞花序顶生，花紫色，花萼钟状。蒴果椭圆形。种子淡褐色，小。花期4～10月，果期8月至翌年4月。

Deciduous shrubs 1~3 m tall; young branchlets, leaves abaxially, petioles, and inflorescences densely rusty pubescent with stellate and/or glandular hairs. Stems brown, branchlets quadrangular to subquadrangular. Leaf blade ovate to elliptic or narrowly, membranous when dry, margin entire to coarsely sinuate-dentate, apex acuminate. Inflorescences terminal, spicate cymes. Corolla purple. Capsules ellipsoid. Seeds pale brown, small. Fl. Apr. - Oct., fr. Aug. - Apr. of following year.

茎干　　Stems
摄影：杨庆松　Photo by: Yang Qingsong

枝叶　　Branch and leaves
摄影：杨庆松　Photo by: Yang Qingsong

花序　　Inflorescense
摄影：杨庆松　Photo by: Yang Qingsong

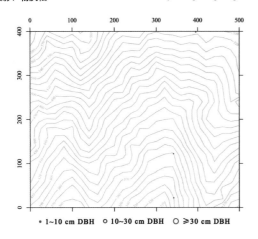

个体分布图　Distribution of individuals

径级分布表　DBH class

胸径区间 (Diameter class) (cm)	个体数 (No. of individuals in the plot)	比例 (Proportion) (%)
1～2	3	100.00
2～5	0	0.00
5～10	0	0.00
10～20	0	0.00
20～30	0	0.00
30～60	0	0.00
≥60	0	0.00

140 厚壳树 Hòu ké shù | Heliotrope Ehretia

Ehretia acuminata R. Brown
紫草科 | Boraginaceae

代码 (SpCode) = EHRACU
个体数 (Individual number/20 hm^2) = 25
最大胸径 (Max DBH) = 34.6 cm
重要值排序 (Importance value rank) = 80

落叶乔木，高达15 m；枝淡褐色，平滑，小枝褐色，无毛，有明显的皮孔。叶椭圆形、倒卵形或长圆状倒卵形，长5～13 cm，宽4～6 cm，先端尖，基部宽楔形，稀圆形，边缘有整齐的锯齿；叶柄长1.5～2.5 cm，无毛。聚伞花序圆锥状；花冠钟状，白色。核果黄色或桔黄色。花期5～6月，果期6～7月。

Deciduous trees to 15 m tall; branches light brown, smooth; branchlets brown, glabrous, with distinct lenticels. Petiole 1.5~2.5 cm, glabrous; leaf blade elliptic to obovate or oblong-obovate, 5~13 × 4~6 cm, base broadly cuneate, margin regularly serrate with teeth curved upward, apex acute, apiculate. Cymes paniculate. Corolla white, campanulate. Drupes yellow or orange. Fl. May - Jun., fr. Jun. - Jul..

树干　Trunk
摄影：杨庆松　Photo by: Yang Qingsong

枝叶　Branch and leaves
摄影：杨庆松　Photo by: Yang Qingsong

花枝　Flowering branch
摄影：汪远　Photo by: Wang Yuan

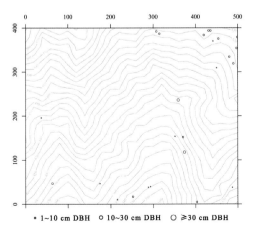
个体分布图 Distribution of individuals

径级分布表 DBH class

胸径等级 (Diameter class) (cm)	个体数 (No. of individuals in the plot)	比例 (Proportion) (%)
1～2	3	12.00
2～5	5	20.00
5～10	1	4.00
10～20	6	24.00
20～30	8	32.00
30～60	2	8.00
≥60	0	0.00

141 华紫珠

Huá zǐ zhū | China Purplepearl

Callicarpa cathayana H. T. Chang
马鞭草科 | Verbenaceae

代码 (SpCode) = CALCAT
个体数 (Individual number/20 hm^2) = 7
最大胸径 (Max DBH) = 3.1 cm
重要值排序 (Importance value rank) = 122

落叶灌木，高1.5～3 m。小枝纤细，幼嫩稍有星状毛，老后脱落。叶片椭圆形或卵形，长4～8 cm，宽1.5～3 cm，顶端渐尖，基部楔形，两面近于无毛，而有显著的红色腺点，侧脉5～7对，在两面均稍隆起，细脉和网脉下陷，边缘密生细锯齿；叶柄长4～8 mm。聚伞花序细弱；花冠紫色。果实球形，紫色，径约2 mm。花期5～7月，果期8～11月。

Deciduous shrubs 1.5~3 m tall. Branchlets slender, slightly stellate tomentose when young, glabrescent. Petiole 4~8 mm; leaf blade elliptic to ovate, 4~8 × 1.5~3 cm, subglabrous, red glandular, base narrowly cuneate, margin serrulate, apex acuminate, veins 5~7 pairs. Cymes slender. Corolla purple. Fruit purple, globose, ca. 2 mm in diam. Fl. May. - Jul., fr. Aug. - Nov..

枝叶   Branch and leaves
摄影：杨庆松   Photo by: Yang Qingsong

果枝   Fruit branch
摄影：杨庆松   Photo by: Yang Qingsong

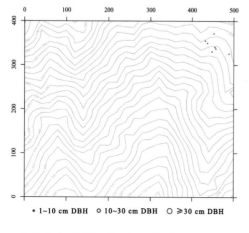

个体分布图 Distribution of individuals

径级分布表 DBH class

胸径区间 (Diameter class) (cm)	个体数 (No. of individuals in the plot)	比例 (Proportion) (%)
1～2	5	71.43
2～5	2	28.57
5～10	0	0.00
10～20	0	0.00
20～30	0	0.00
30～60	0	0.00
≥60	0	0.00

142 老鸦糊

Lǎo yā hú | Girald Purplepearl

Callicarpa giraldii Hesse ex Rehd.
马鞭草科 | Verbenaceae

代码 (SpCode) = CALGIR
个体数 (Individual number/20 hm^2) = 30
最大胸径 (Max DBH) = 4.2 cm
重要值排序 (Importance value rank) = 103

落叶灌木，高1～3 (5) m。小枝圆柱形，被星状毛。叶片纸质，宽椭圆形至披针状长圆形，顶端渐尖，基部楔形或下延成狭楔形，边缘有锯齿，表面黄绿色，稍有微毛，背面淡绿色，疏被星状毛和细小黄色腺点。聚伞花序，具星状毛；花紫色。果实球形，紫色。花期5～6月，果期7～11月。

Deciduous shrubs 1~3(~5) m tall. Branches terete, stellate tomentose; leaf blade lanceolate, oblong, oblong-elliptic, elliptic, broadly elliptic, obovate-oblong, or broadly ovate, papery, base cuneate, narrowly cuneate, obtuse, or rounded, margin serrate, apex acuminate. Cymes, stellate tomentose. Corolla purple. Fruit purple. Fl. May - Jul., fr. Jul. - Nov..

枝叶　Branch and leaves
摄影：杨庆松　Photo by: Yang Qingsong

果枝　Fruiting branches
摄影：杨庆松　Photo by: Yang Qingsong

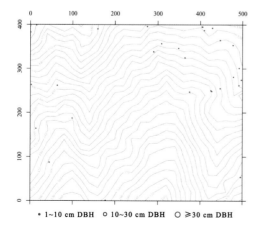

个体分布图 Distribution of individuals

径级分布表 DBH class

胸径等级 (Diameter class) (cm)	个体数 (No. of individuals in the plot)	比例 (Proportion) (%)
1～2	14	46.67
2～5	16	53.33
5～10	0	0.00
10～20	0	0.00
20～30	0	0.00
30～60	0	0.00
≥60	0	0.00

143 秃红紫珠

Tū hóng zǐ zhū | Subglabrous Reddish Purplepearl

Callicarpa rubella var. *subglabra* (Pei) H. T. Chang
马鞭草科 | Verbenaceae

代码 (SpCode) = CALRUB
个体数 (Individual number/20 hm^2) = 3
最大胸径 (Max DBH) = 2.8 cm
重要值排序 (Importance value rank) = 134

落叶灌木，高约2 m；小枝、叶片、花序和花萼、花冠均无毛。叶片倒卵形或倒卵状椭圆形，顶端尾尖或渐尖，基部心形，有时偏斜，边缘具细锯齿或不整齐的粗齿；叶柄极短或近于无柄。聚伞花序宽2～4 cm；花冠紫红色、黄绿色或白色。果实紫红色。花期5～7月，果期7～11月。

Deciduous shrubs 2 m tall. Branchlets, leaves, inflorescences, calyces, and corollas glabrous. Petiole very short to leaf subsessile; leaf blade lanceolate, oblanceolate, obovate-elliptic, or ovate, base cordate and sometimes oblique, margin serrulate, irregularly serrate, or serrate, apex acuminate to caudate. Cymes 2~4 cm across. Corolla purple, greenish, or white. Fruit purple. Fl. May - Jul., fr. Jul. - Nov..

枝叶　　Branch and leaves
摄影：杨庆松　　Photo by: Yang Qingsong

果枝　　Fruiting Branch
摄影：杨庆松　　Photo by: Yang Qingsong

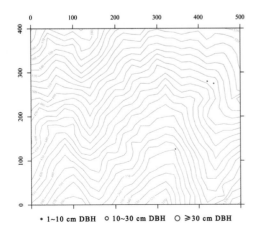

个体分布图　Distribution of individuals

径级分布表 DBH class

胸径区间 (Diameter class) (cm)	个体数 (No. of individuals in the plot)	比例 (Proportion) (%)
1～2	1	33.33
2～5	2	66.67
5～10	0	0.00
10～20	0	0.00
20～30	0	0.00
30～60	0	0.00
≥60	0	0.00

144 灰毛大青

Huī máo dà qīng | Greyhair Glorybower

Clerodendrum canescens Wallich ex Walpers
马鞭草科 | Verbenaceae

代码 (SpCode) = CLECAN
个体数 (Individual number/20 hm^2) = 22
最大胸径 (Max DBH) = 3.8 cm
重要值排序 (Importance value rank) = 106

落叶灌木，高1~3.5 m；小枝略四棱形，全体密被平展或倒向灰褐色长柔毛。叶片心形或宽卵形，少为卵形，长6~18 cm，宽4~15 cm，顶端渐尖，基部心形至近截形。聚伞花序密集成头状，花冠白色或淡红色。核果近球形，径约7 mm，绿色，成熟时深蓝色或黑色，藏于红色增大的宿萼内。花果期4~10月。

Deciduous shrubs 1~3.5 m tall. Branchlets 4-angled. Whole plant densely yellowish to brownish tomentose to nearly villous. Leaf blade broadly ovate or cordate, rarely ovate, base cordate to subtruncate, apex acuminate. Inflorescences terminal, subconical. Corolla white or pinkish. Drupes green when young, dark blue to black at maturity, subglobose, ca. 7 mm in diam. Fl. and fr. Apr. - Oct..

幼叶　　Young leaves
摄影：雷霄　Photo by: Lei Xiao

花枝　　Flowering branch
摄影：王樟华　Photo by: Wang Zhanghua

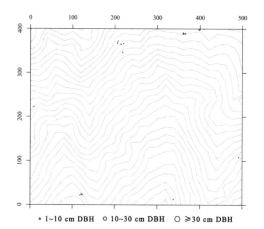

个体分布图　Distribution of individuals

径级分布表 DBH class

胸径等级 (Diameter class) (cm)	个体数 (No. of individuals in the plot)	比例 (Proportion) (%)
1~2	13	59.09
2~5	9	40.91
5~10	0	0.00
10~20	0	0.00
20~30	0	0.00
30~60	0	0.00
≥60	0	0.00

145 大青   Dà qīng | Manyflower Glorybower

Clerodendrum cyrtophyllum Turcz.
马钱科 | Loganiaceae

代码 (SpCode) = CLECYR
个体数 (Individual number/20 hm^2) = 429
最大胸径 (Max DBH) = 6.7 cm
重要值排序 (Importance value rank) = 43

落叶灌木；幼枝被短柔毛，枝黄褐色。叶片纸质，椭圆形、卵状椭圆形、长圆形或长圆状披针形，顶端渐尖或急尖，基部圆形或宽楔形，通常全缘，两面无毛或沿脉疏生短柔毛，背面常有腺点。伞房状聚伞花序，生于枝顶或叶腋；花冠白色。果实球形或倒卵形，蓝紫色。花果期6月至翌年2月。

Deciduous shrubs. Branchlets yellow-brown, pubescent. Leaf blade oblong, oblong-lanceolate, elliptic, or ovate-elliptic, papery, glabrous or sparsely pubescent along veins, abaxially glandular, base rounded to cuneate, margin entire or rounded serrate, apex acuminate to acute. Inflorescences corymbose, formed from terminal and subterminal cymes. Corolla white. Drupes blue-purple, obovate to globose. Fl. and fr. Jun. - Feb. of following year.

花枝   Flowering branch
摄影：杨庆松   Photo by: Yang Qingsong

枝叶   Branch and leaves
摄影：杨庆松   Photo by: Yang Qingsong

果枝   Fruiting Branch
摄影：杨庆松   Photo by: Yang Qingsong

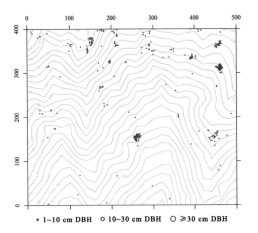

个体分布图 Distribution of individuals

径级分布表 DBH class

胸径区间 (Diameter class) (cm)	个体数 (No. of individuals in the plot)	比例 (Proportion) (%)
1~2	253	58.97
2~5	169	39.39
5~10	7	1.63
10~20	0	0.00
20~30	0	0.00
30~60	0	0.00
≥60	0	0.00

146 豆腐柴

Dòu fu chái | Japan Premna

Premna microphylla Turcz.
马鞭草科 | Verbenaceae

代码 (SpCode) = PREMIC
个体数 (Individual number/20 hm^2) = 20
最大胸径 (Max DBH) = 4.1 cm
重要值排序 (Importance value rank) = 108

直立落叶灌木。幼枝有柔毛，老枝变无毛。叶揉之有臭味，卵状披针形，椭圆形，卵形，或倒卵形，基部狭楔形，边缘全缘或浅裂有时有细锯齿，先端长渐尖或急尖。圆锥花序，花萼绿色，略带紫色，杯状，花冠淡黄。果紫色，球状至倒卵形。花期5～6月，果期8～10月。

Deciduous shrubs, erect. Branchlets pubescent when young, glabrescent. The smell of leaf when rubing; leaf blade ovate-lanceolate, elliptic, ovate, or obovate, base narrowly cuneate, margin entire or lobed to sometimes serrulate, apex long acuminate to acute. Inflorescences conical panicles. Calyx green and slightly purplish, cup-shaped, corolla yellowish,. Fruit purple, globose to obovate. Fl. May - Jun., fr. Aug. - Oct..

枝叶　Branch and leaves
摄影：杨庆松　Photo by: Yang Qingsong

老叶　Old leaves
摄影：杨庆松　Photo by: Yang Qingsong

花枝　Flowering branch
摄影：杨庆松　Photo by: Yang Qingsong

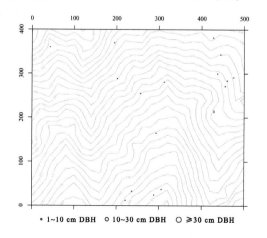

个体分布图　Distribution of individuals

径级分布表　DBH class

胸径等级 (Diameter class) (cm)	个体数 (No. of individuals in the plot)	比例 (Proportion) (%)
1～2	11	55.00
2～5	9	45.00
5～10	0	0.00
10～20	0	0.00
20～30	0	0.00
30～60	0	0.00
≥60	0	0.00

147 茜树

Qiàn shù | Maddertree

Aidia cochinchinensis Lour.
茜草科 | Rubiaceae

代码 (SpCode) = AIDCOC
个体数 (Individual number/20 hm^2) = 4
最大胸径 (Max DBH) = 1.8 cm
重要值排序 (Importance value rank) = 132

常绿灌木；枝无毛。叶革质或纸质，对生，椭圆形至披针形。聚伞花序与叶对生或生于无叶的节上，多花，花冠黄色或白色。浆果球形，紫黑色，种子多数。花期3～6月，果期5月至翌年2月。

Evergreen shrubs; branches glabrous. Leaf blade drying leathery or papery, opposite, elliptic to lanceolate. Inflorescences cymose, corolla yellow or white, many flowers. Berry globose, purple-black, many seeds. Fl. Mar. - Jun., fr. May - Feb. of following year.

树干　　Trunk
摄影：杨庆松　Photo by: Yang Qingsong

枝叶　　Branch and leaves
摄影：杨庆松　Photo by: Yang Qingsong

果枝　　Fruiting Branch
摄影：杨庆松　Photo by: Yang Qingsong

径级分布表 DBH class

胸径区间 (Diameter class) (cm)	个体数 (No. of individuals in the plot)	比例 (Proportion) (%)
1～2	4	100.00
2～5	0	0.00
5～10	0	0.00
10～20	0	0.00
20～30	0	0.00
30～60	0	0.00
≥60	0	0.00

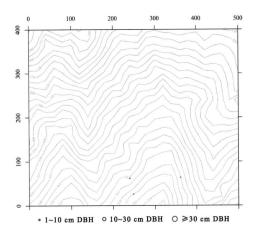

个体分布图 Distribution of individuals

148 狗骨柴

Gǒu gǔ chái | Common Dogbonebavin

Diplospora dubia (Lindl.) Masam.
茜草科 | Rubiaceae

代码 (SpCode) = DIPDUB
个体数 (Individual number/20 hm^2) = 1
最大胸径 (Max DBH) = 1.8 cm
重要值排序 (Importance value rank) = 152

常绿灌木或小乔木，高1～12 m。一年生枝灰黄色，光滑无毛。单叶对生，叶片近革质，全缘，叶面具光泽，两面无毛。伞房状聚伞花序腋生，总花梗极短，花冠绿白色。浆果近球形，橙红色，干后黑色。花期5～6月，果期7～10月。

Everygreen shrubs or small trees, to 1~12 m tall. First year branches greyish yellow, glabrous. Simple leaves opposite, blade nearly leathery, margin entire, shiny, glabrous on both surfaces. Inforescence compoundly cyme, axillary, peduncle very short, corolla green and white. Berries subglobose, orange, dry fruit black. Fl. May - Jun., fr. Jul. - Oct..

枝叶　Branch and leaves
摄影：杨庆松　Photo by: Yang Qingsong

果枝　Fruiting branches
摄影：杨庆松　Photo by: Yang Qingsong

果　Fruits
摄影：杨庆松　Photo by: Yang Qingsong

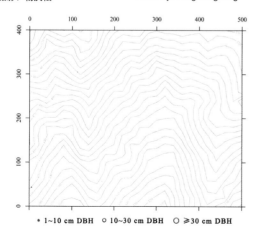

个体分布图　Distribution of individuals

径级分布表 DBH class

胸径等级 (Diameter class) (cm)	个体数 (No. of individuals in the plot)	比例 (Proportion) (%)
1～2	1	100.00
2～5	0	0.00
5～10	0	0.00
10～20	0	0.00
20～30	0	0.00
30～60	0	0.00
≥60	0	0.00

149 栀子   Zhī zi | Cape Jasmine

Gardenia jasminoides Ellis
茜草科 | Rubiaceae

代码 (SpCode) = GARJAS
个体数 (Individual number/20 hm^2) = 1
最大胸径 (Max DBH) = 3.4 cm
重要值排序 (Importance value rank) = 147

常绿灌木，高0.3～3 m；嫩枝常被短毛，枝圆柱形，灰色。叶对生，革质，稀为纸质。花芳香，通常单朵生于枝顶，花冠白色或乳黄色。种子多数，扁平，近圆形而稍有棱角。花期3～7月，果期5月至翌年2月。

Evergreen shrubs, 0.3~3 m tall. Young branches usually densely puberulent, terete, gray. Leaves opposite or rarely ternate, leathery to stiffly papery. Flowers are fragrant, usually single on the top of the branch, corolla white or milky yellow. Seeds are mostly flat, nearly round and slightly angular. Fl. Mar. - Jul., fr. May - Feb. of following year.

枝叶   Branch and leaves
摄影：杨庆松   Photo by: Yang Qingsong

叶   Leaves
摄影：杨庆松   Photo by: Yang Qingsong

花   Flower
摄影：杨庆松   Photo by: Yang Qingsong

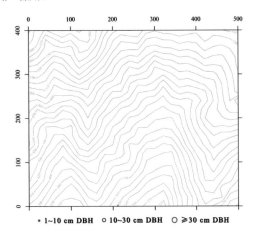
个体分布图   Distribution of individuals

径级分布表 DBH class

胸径区间 (Diameter class) (cm)	个体数 (No. of individuals in the plot)	比例 (Proportion) (%)
1～2	0	0.00
2～5	1	100.00
5～10	0	0.00
10～20	0	0.00
20～30	0	0.00
30～60	0	0.00
≥60	0	0.00

150 藕花 (大叶白纸扇)　　　　Lí huā | Esquirol Jadeleaf and Goldenflower

Mussaenda shikokiana Makino
茜草科 | Rubiaceae

代码 (SpCode) = MUSSHI
个体数 (Individual number/20 hm^2) = 36
最大胸径 (Max DBH) = 4.1 cm
重要值排序 (Importance value rank) = 101

直立或攀援落叶灌木，高1～3 m。枝圆柱状，密被短糙伏毛。叶对生；叶柄1.5～3.5 cm，有毛；叶片薄纸质，正面绿色到淡绿色，背面浅灰色到白色。花序头状聚伞花序，萼裂片近叶状，白色，披针形至舌状。浆果近球形，直径约1 cm。花期5～7月，果期7～10月。

Erect or climbing deciduous shrubs, 1~3 m tall; branches terete, densely strigillose. Leaves opposite; petiole 1.5~3.5 cm, with strigillose; blade drying thinly papery, adaxially green to pale green, abaxially pale gray to whitened. Inflorescences subcapitate, lobes subleaflike, white, lanceolate to ligulate. Berry subglobose to ellipsoid, ca. 1 cm. Fl. May - Jul., fr. Jul. - Oct..

花枝　　　　Flowering branch
摄影：王樟华　Photo by: Wang Zhanghua

枝叶　　　　Branch and leaves
摄影：杨庆松　Photo by: Yang Qingsong

果枝　　　　Fruiting branch
摄影：杨庆松　Photo by: Yang Qingsong

径级分布表 DBH class

胸径等级 (Diameter class) (cm)	个体数 (No. of individuals in the plot)	比例 (Proportion) (%)
1～2	25	69.44
2～5	11	30.56
5～10	0	0.00
10～20	0	0.00
20～30	0	0.00
30～60	0	0.00
≥60	0	0.00

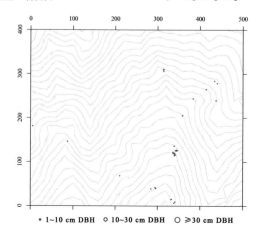

个体分布图 Distribution of individuals

151 鸡仔木

Jī zǎi mù | Chjickenwood

Sinoadina racemosa (Sieb. et Zucc.) Ridsd.
茜草科 | Rubiaceae

代码 (SpCode) = SINRAC
个体数 (Individual number/20 hm^2) = 11
最大胸径 (Max DBH) = 20.4 cm
重要值排序 (Importance value rank) = 104

落叶乔木，高4~12 m；树皮灰色，小枝无毛。叶对生，薄革质，基部心形或钝，有时偏斜。叶柄长3~6 cm，无毛或有短柔毛。头状花序，花冠淡黄色；小蒴果倒卵状楔形。花、果期5~12月。

Deciduous trees, 4~12 m tall; trunk bark gray; branches glabrous. Leaf opposite, leaf blade drying thinly leathery, base cordate to obtuse, sometimes slightly inequilateral. Petiole 3~6 cm, glabrous or puberulent. Capitate cymes, corolla light yellow. Capsules obovoid-cuneate. Fl. and fr. May - Dec..

树干　Trunk
摄影：杨庆松　Photo by: Yang Qingsong

叶　Leaves
摄影：杨庆松　Photo by: Yang Qingsong

小枝　Branch
摄影：杨庆松　Photo by: Yang Qingsong

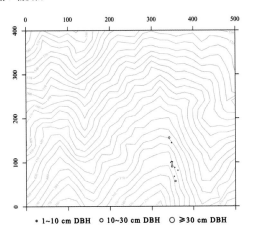

个体分布图　Distribution of individuals

径级分布表　DBH class

胸径区间 (Diameter class) (cm)	个体数 (No. of individuals in the plot)	比例 (Proportion) (%)
1~2	0	0.00
2~5	3	27.27
5~10	5	45.45
10~20	2	18.18
20~30	1	9.09
30~60	0	0.00
≥60	0	0.00

152 白花苦灯笼

Bái huā kǔ dēng long | Whiteflower Tarenna

Tarenna mollissima (Hook. et Arn.) B. L. Robinson
茜草科 | Rubiaceae

代码 (SpCode) = TARMOL
个体数 (Individual number/20 hm^2) = 14
最大胸径 (Max DBH) = 2.1 cm
重要值排序 (Importance value rank) = 115

落叶灌木，高1～3 m，全株密被灰色或褐色柔毛或短绒毛，但老枝毛渐脱落。叶纸质，披针形、长圆状披针形或卵状椭圆形。聚伞花序顶生，多花；花冠白色。浆果近球形，直径5～7 mm。花期5～7月，果期5月至翌年2月。

Deciduous shrubs, 1~3 m tall; branches densely gray or brown pilosulous or tomentulose, becoming glabrescent when old. Leaf blade drying papery and blackish brown, lanceolate, oblong-lanceolate, or ovate-elliptic. Inflorescences corymbose, many flowered, corolla white. Berry subglobose, 5~7 mm in diam. Fl. May - Jul., fr. May - Feb. of following year.

叶　Leaves
摄影：王樟华　Photo by: Wang Zhanghua

花序　Inflorescence
摄影：杨庆松　Photo by: Yang Qingsong

果枝　Fruiting branch
摄影：杨庆松　Photo by: Yang Qingsong

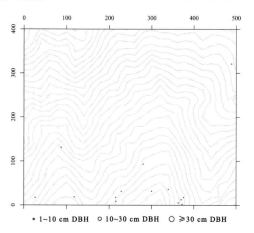
个体分布图　Distribution of individuals

径级分布表 DBH class

胸径等级 (Diameter class) (cm)	个体数 (No. of individuals in the plot)	比例 (Proportion) (%)
1～2	12	85.71
2～5	2	14.29
5～10	0	0.00
10～20	0	0.00
20～30	0	0.00
30～60	0	0.00
≥60	0	0.00

153 宜昌荚蒾

Yí chāng jiá mí | Yichang Arrowwood

Viburnum erosum Thunb.
忍冬科 | Caprifoliaceae

代码 (SpCode) = VIBERO
个体数 (Individual number/20 hm^2) = 103
最大胸径 (Max DBH) = 4.1 cm
重要值排序 (Importance value rank) = 83

落叶灌木，高可达3 m。当年小枝连同芽、叶柄和花序均密被簇状和简单长柔毛。单叶对生，叶缘有波状小尖齿，有托叶。复伞形状聚伞花序。浆果状核果宽卵圆形，红色。花期4~5月，果期9~10月。

Deciduous shrubs, up to 3 m tall. First year branchlet, bud, petiole, inflorescence and calyx with dense satellite and simple villus. Simple leaves opposite, with small sharp teeth on margins, with stipules. Inflorescence compoundly cyme. Berry-like drupe broad ovate, red at maturity. Fl. Apr. - May, fr. Sep. - Oct..

果枝　　Fruiting branch
摄影：杨庆松　Photo by: Yang Qingsong

花枝　　Flowering branch
摄影：杨庆松　Photo by: Yang Qingsong

果　　Fruits
摄影：杨庆松　Photo by: Yang Qingsong

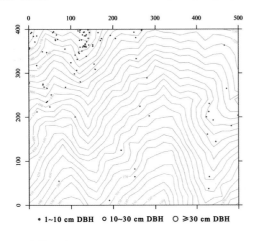

个体分布图　Distribution of individuals

径级分布表　DBH class

胸径区间 (Diameter class) (cm)	个体数 (No. of individuals in the plot)	比例 (Proportion) (%)
1~2	76	73.79
2~5	27	26.21
5~10	0	0.00
10~20	0	0.00
20~30	0	0.00
30~60	0	0.00
≥60	0	0.00

154 棕榈　　　　　　　　　　　　　　　　　　　　　　　Zōng lǘ | Palm

Trachycarpus fortunei (Hook.) H. Wendl.
棕榈科 | Palmae

代码 (SpCode) = TRAFOR
个体数 (Individual number/20 hm²) = 1
最大胸径 (Max DBH) = 10.2 cm
重要值排序 (Importance value rank) = 140

常绿植物，乔木状，高可达10 m或更高，树干直径10~15 cm甚至更粗。叶片呈3/4圆形或者近圆形。花序粗壮，多次分枝，从叶腋抽出，通常是雌雄异株。果实阔肾形，成熟时由黄色变为淡蓝色，有白粉。花期4月，果期12月。

Trees-shaped, evergreen, up to 10 m tall or more. Trunk diameter up to 10~15 cm or thicker. Leaves are 3/4 round or nearly round. The inflorescence stout, multiple branches and from taking the axil, usually dioecious. Fruits kidney-shaped, mature from yellow to pale blue, with white powder. Fl. Apr., fr. Dec..

花序　　Inflorescence
摄影：王樟华　Photo by: Wang Zhanghua

枝叶　　Branch and leaves
摄影：杨庆松　Photo by: Yang Qingsong

树冠　　Canopy
摄影：王樟华　Photo by: Wang Zhanghua

径级分布表 DBH class

胸径等级 (Diameter class) (cm)	个体数 (No. of individuals in the plot)	比例 (Proportion) (%)
1~2	0	0.00
2~5	0	0.00
5~10	0	0.00
10~20	1	100.00
20~30	0	0.00
30~60	0	0.00
≥60	0	0.00

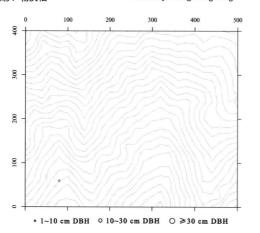

个体分布图 Distribution of individuals

附录I 植物中文名索引
Appendix I Chinese Species Name Index

A
矮小天仙果 ················ 40

B
白背叶 ···················· 80
白花苦灯笼 ················ 163
白檀 ······················ 140
百齿卫矛 ·················· 96
薄叶润楠 ·················· 56
薄叶山矾 ·················· 135
豹皮樟 ···················· 53
笔罗子 ···················· 104

C
糙叶树 ···················· 33
茶 ·························· 112
檫木 ······················ 60
赤楠 ······················ 122
赤皮青冈 ·················· 25
赤杨叶 ···················· 143
臭辣树 ···················· 77
刺毛越桔 ·················· 132
刺楸 ······················ 125
刺叶桂樱 ·················· 70
长叶冻绿 ·················· 106

D
大青 ······················ 156
大叶白纸扇 ················ 161
大叶冬青 ·················· 93
大叶榉树 ·················· 39
大叶早樱 ·················· 67
灯台树 ···················· 126
冬青 ······················ 90
豆腐柴 ···················· 157
杜鹃 ······················ 129
杜英 ······················ 108
短梗冬青 ·················· 89

F
榧树 ······················ 15
枫香树 ···················· 64

G
港柯 ······················ 32
格药柃 ···················· 115
狗骨柴 ···················· 159
构棘 ······················ 41
拐枣 ······················ 105
光亮山矾 ·················· 139
光叶山矾 ·················· 138
光叶石楠 ·················· 72

H
海金子 ···················· 61
海桐山矾 ·················· 137
杭州榆 ···················· 38
褐叶青冈 ·················· 30
黑山山矾 ·················· 137
红淡比 ···················· 113
红毒茴 ···················· 47
红果钓樟 ·················· 50
红果山胡椒 ················ 50
红脉钓樟 ·················· 52
红楠 ······················ 57
红皮树 ···················· 145
红枝柴 ···················· 103
厚壳树 ···················· 151
厚皮香 ···················· 118
胡桃楸 ···················· 19
胡颓子 ···················· 120
湖北山楂 ·················· 68
湖北算盘子 ················ 79
虎皮楠 ···················· 84
华东木犀 ·················· 148
华东楠 ···················· 56
华东野核桃 ················ 19
华紫珠 ···················· 152
化香树 ···················· 20
黄丹木姜子 ················ 55
黄毛楤木 ·················· 123
黄牛奶树 ·················· 136
黄檀 ······················ 76
灰毛大青 ·················· 155

J
鸡仔木 ···················· 162
棘茎楤木 ·················· 124
檵木 ······················ 65
江南越桔 ·················· 131

K
栲树 ······················ 23
柯 ·························· 31
苦枥木 ···················· 146
苦槠 ······················ 24

L
老鼠矢 ···················· 141
老鸦糊 ···················· 153
雷公鹅耳枥 ················ 21
䕨花 ······················ 161
楝叶吴萸 ·················· 77

M
马银花 ···················· 128
毛八角枫 ·················· 121
毛花连蕊茶 ················ 110
毛脉槭 ···················· 101
米槠 ······················ 22
密果吴萸 ·················· 78
木荷 ······················ 117
木蜡树 ···················· 88
木犀 ······················ 149

N
南京椴 ···················· 109
南酸枣 ···················· 85
南天竹 ···················· 45
南烛 ······················ 130
拟赤杨 ···················· 143
宁波木犀 ·················· 148
牛鼻栓 ···················· 63

P
披针叶茴香 ················ 47
披针叶山矾 ················ 138
朴树 ······················ 35

Q

茜树	158
青冈	26
青灰叶下珠	82
青栲	28
青皮木	44
青钱柳	18

R

锐角槭	99

S

赛山梅	144
三尖杉	14
三角槭	100
山苍子	54
山茶	111
山矾	142
山合欢	75
山胡椒	51
山槐	75
山鸡椒	54
山柿	133
山桐子	119
山油麻	37
杉木	13
石斑木	74
石栎	31
栓叶安息香	145
四川山矾	139
四照花	127

T

天目木兰	46
铁冬青	95
秃红紫珠	154

W

蕨芝	41
乌饭树	130
无患子	102
吴茱萸	78

X

西川朴	36
细刺枸骨	91
细叶青冈	27
细枝柃	114
腺叶桂樱	69
香桂	49
响叶杨	16
小果冬青	94
小蜡	147
小叶青冈	28
小叶石楠	73

Y

崖花海桐	61
盐肤木	86
杨梅	17
杨梅叶蚊母树	62
杨桐	113
野漆	87
野桐	81
野鸦椿	98
宜昌荚蒾	164
银杏	12
迎春樱桃	66
油柿	134
油桐	83
云山青冈	29

Z

窄基红褐柃	116
樟	48
柘	42
浙江新木姜子	58
栀子	160
枳椇	105
中华杜英	107
中华石楠	71
中华卫矛	97
皱柄冬青	92
紫弹树	34
紫麻	43
紫楠	59
棕榈	165
醉鱼草	150

附录II　植物拉丁名索引
Appendix II　Scientific Species Name Index

A

Acer acutum	99
Acer buergerianum	100
Acer pubinerve	101
Aidia cochinchinensis	158
Alangium kurzii	121
Albizia kalkora	75
Alniphyllum fortunei	143
Aphananthe aspera	33
Aralia chinensis	123
Aralia echinocaulis	124

B

Buddleja lindleyana	150

C

Callicarpa cathayana	152
Callicarpa giraldii	153
Callicarpa rubella	154
Camellia fraterna	110
Camellia japonica	111
Camellia sinensis	112
Carpinus viminea	21
Castanopsis carlesii	22
Castanopsis fargesii	23
Castanopsis sclerophylla	24
Celtis biondii	34
Celtis sinensis	35
Celtis vandervoetiana	36
Cephalotaxus fortunei	14
Cerasus discoidea	66
Cerasus subhirtella	67
Choerospondias axillaris	85
Cinnamomum camphora	48
Cinnamomum subavenium	49
Clerodendrum canescens	155
Clerodendrum cyrtophyllum	156
Cleyera japonica	113
Cornus controversa	126
Cornus kousa subsp. *chinensis*	127
Crataegus hupehensis	68
Cunninghamia lanceolata	13
Cyclobalanopsis gilva	25
Cyclobalanopsis glauca	26
Cyclobalanopsis gracilis	27
Cyclobalanopsis myrsinifolia	28
Cyclobalanopsis sessilifolia	29
Cyclobalanopsis stewardiana	30
Cyclocarya paliurus	18

D

Dalbergia hupeana	76
Daphniphyllum oldhami	84
Diospyros japonica	133
Diospyros oleifera	134
Diplospora dubia	159
Distylium myricoides	62

E

Ehretia acuminata	151
Elaeagnus pungens	120
Elaeocarpus chinensis	107
Elaeocarpus decipiens	108
Euonymus centidens	96
Euonymus nitidus	97
Eurya loquaiana	114
Eurya muricata	115
Eurya rubiginosa	116
Euscaphis japonica	98

F

Ficus erecta	40
Fortunearia sinensis	63
Fraxinus insularis	146

G

Gardenia jasminoides	160
Ginkgo biloba	12
Glochidion wilsonii	79

H

Hovenia acerba	105

I
Idesia polycarpa ... 119
Ilex buergeri ... 89
Ilex chinensis ... 90
Ilex hylonoma ... 91
Ilex kengii ... 92
Ilex latifolia ... 93
Ilex micrococca ... 94
Ilex rotunda ... 95
Illicium lanceolatum ... 47

J
Juglans mandshurica ... 19

K
Kalopanax septemlobus ... 125

L
Laurocerasus phaeosticta ... 69
Laurocerasus spinulosa ... 70
Ligustrum sinense ... 147
Lindera erythrocarpa ... 50
Lindera glauca ... 51
Lindera rubronervia ... 52
Liquidambar formosana ... 64
Lithocarpus glaber ... 31
Lithocarpus harlandii ... 32
Litsea coreana var. *sinensis* ... 53
Litsea cubeba ... 54
Litsea elongata ... 55
Loropetalum chinense ... 65

M
Machilus leptophylla ... 56
Machilus thunbergii ... 57
Maclura cochinchinensis ... 41
Maclura tricuspidata ... 42
Magnolia amoena ... 46
Mallotus apelta ... 80
Mallotus tenuifolius ... 81
Meliosma oldhamii ... 103
Meliosma rigida ... 104
Mussaenda shikokiana ... 161
Myrica rubra ... 17

N
Nandina domestica ... 45
Neolitsea aurata var. *chekiangensis* ... 58

O
Oreocnide frutescens ... 43
Osmanthus cooperi ... 148
Osmanthus fragrans ... 149

P
Phoebe sheareri ... 59
Photinia beauverdiana ... 71
Photinia glabra ... 72
Photinia parvifolia ... 73
Phyllanthus glaucus ... 82
Pittosporum illicioides ... 61
Platycarya strobilacea ... 20
Populus adenopoda ... 16
Premna microphylla ... 157

R
Rhamnus crenata ... 106
Rhaphiolepis indica ... 74
Rhododendron ovatum ... 128
Rhododendron simsii ... 129
Rhus chinensis ... 86

S
Sapindus saponaria ... 102
Sassafras tzumu ... 60
Schima superba ... 117
Schoepfia jasminodora ... 44
Sinoadina racemosa ... 162
Styrax confusus ... 144
Styrax suberifolius ... 145
Symplocos anomala ... 135
Symplocos cochinchinensis ... 136
Symplocos heishanensis ... 137
Symplocos lancifolia ... 138
Symplocos lucida ... 139
Symplocos paniculata ... 140
Symplocos stellaris ... 141
Symplocos sumuntia ... 142
Syzygium buxifolium ... 122

T

Tarenna mollissima ···163
Ternstroemia gymnanthera ···································118
Tetradium glabrifolium ···77
Tetradium ruticarpum ···78
Tilia miqueliana ··109
Torreya grandis ··15
Toxicodendron succedaneum ·································87
Toxicodendron sylvestre ··88
Trachycarpus fortunei ···165
Trema cannabina var. *dielsiana* ·····························37

U

Ulmus changii ··38

V

Vaccinium bracteatum ··130
Vaccinium mandarinorum ····································131
Vaccinium trichocladum ······································132
Vernicia fordii ··83
Viburnum erosum ··164

Z

Zelkova schneideriana ···39